OLD MAN MUSCLE

Old Man Muscle
Strength Training Handbook

Peter S. Murano

ISBN 978-1-7340764-1-7

This book is a work of non-fiction and both author and publisher make no explicit guarantee as to the accuracy of the information it contains. Web addresses that appear in the Resources sections of this book are volatile due to the ever-changing nature of the internet. The reader of this book should be aware that it is possible that some web links may become invalid after this work has been published.

Persons appearing in photographs within the pages of this book were either derived from royalty-free iStock (Getty) images created expressly for publication, or were generated by the author using any live models of his choosing.

Printed in the United States of America

Table of Contents

NOTICE

The information, including but not limited to, text, graphics, images, and other material contained in this publication are for informational purposes only. The purpose of this handbook is to promote broad consumer understanding and knowledge of various health and fitness topics. It is not intended to be a substitute for professional medical advice, diagnosis or treatment. Always seek the advice of your physician or other qualified health care provider with any questions you may have regarding a medical condition or treatment, and prior to undertaking a new health care regimen. Never disregard professional medical advice or delay in seeking it because of something you have read in this handbook.

Dr. Peter S. Murano/Old Man Muscle™ do not recommend or endorse any specific tests, physicians, products, procedures, opinions or other information that may be mentioned in this publication. Reliance on any information appearing in this handbook is solely at your own risk.

Dr. Peter S. Murano/Old Man Muscle™ do not endorse, nor have any contractual relationships with, any specific health products or supplements. We provide information based upon professional credentials and expertise, and personal experience regarding dietary/nutritional advice, resistance training and muscle recovery strategies, and improving overall well-being, for men 50+ years old, as they age.

Peter S. Murano
M.S., Ph.D. Human Nutrition and Foods, Virginia Tech, 1986, 1989
Certified Personal Trainer, ACE (American Council on Exercise) 2019
Certified Fitness Trainer, ISSA (International Sports Sciences Association) 2019

ACKNOWLEDGMENTS

First, as befits a book dealing with strength training, I wish to acknowledge the strongest person that I know of. Curiously it is not a man, nor is it an individual of imposing physical strength. It is my five-foot tall wife, Dr. Elsa Murano. Despite her diminutive stature, she is a true tower of strength, and lives a life displaying leadership, intelligence, good character, compassion, and integrity to levels that I have never seen equaled - by anyone. And, she has supported me all the way as my passion for strength training has grown. I dedicate this book to her, and to my brother Robert, and my extended family and friends.

There are several colleagues and individuals in the fitness field, some of whom I have worked with, and others I have yet to meet, who have also influenced me and are reasons that this handbook exists. These include Monica Latham, Terri Garcia, Terri Elliot, Dr. Len Kravitz, Dr. Richard Kreider, Dr. Steven Riechman, Dr. Brad Schoenfeld, Dr. Frank Thomas, Dr. Darryl Willoughby, Dr. Yuxiang Sun, Brad Borland, Pete Koch, Jeff Cavalier, Gerald Brainum, and Ric Drasin.

Special thanks to fitness model Chad James and to photographer Skylar Jordan Brown (Seyle Daniel Photography) for the Old Man Muscle workout photos; also to photographer Toni Aull (colorblindxs2photography) for additional photos.

And, I wish to acknowledge the One who has breathed life into each one of us, the Lord God Almighty, creating our physical bodies as well as our souls.

I will praise thee; for I am fearfully and wonderfully made: marvelous are thy works; and that my soul knoweth right well. (Psalm 139:14)

INTRODUCTION

Greetings and welcome to my Old Man Muscle Handbook on strength training. I want to thank you for picking up this book and reading it. What it tells me is that you are motivated to change your life – for the better. And that is something very positive, and commendable. In this initial chapter, I am going to introduce myself and tell the reader why I wrote this particular book.

My name is Peter Murano, and I started writing this handbook in 2019 as a 65 year-old man living in Texas with my wife, Elsa Murano. I am a college professor at Texas A&M University, specifically a member of the faculty of Nutrition and Food Science.

I was born in Amityville, New York back in 1954, to Harry and Jean Murano. My dad, a native of Scranton, PA whose parents were from Italy, was stationed in Great Britain during WWII in the U.S. Army, and met my mom there and was married to her by mom's father, a Welshman and Baptist minister named Gwillym Oswald Griffith. I have an older brother, Robert. That's it – a small family.

I grew up on Long Island, New York and lived there from 1954 until my college years. Life moved me out of my home state and I took up residence in Florida, Virginia, Iowa, and eventually Texas (in 1995). I have lived more years in the Lone Star State than elsewhere, so I consider myself a Texan. Wasn't born here, but "got here as fast as I could", as a bumper sticker says.

Growing up, I actually yearned to be a Texan, because I had a great love of Texas history, and in particular, the battle of the Alamo. I was a big fan of the Fess Parker and John Wayne screen portrayals of Davy Crockett, and when I actually moved to Texas, I quickly made my way to San Antonio to pay my respects. Have been back to that pretty city countless times; I never get tired of visiting the Alamo.

As kids, my brother and I were fans of the NY Yankees, who back then had the likes of Mickey Mantle, Roger Maris, Whitey Ford, Joe Pepitone, Tom Tresh and Ellie Howard on the team. Phil Rizzuto, Joe Garagiola, and Red Barber would announce the Yankee games on radio and television. We loved baseball through our early teens, but later by the time we got to high school, our attention had shifted to basketball.

This is the time when "Pistol Pete" Maravich was attending LSU, soon to turn pro in the NBA. He was and still is one of my heroes. The NY Knicks had the great players and championship teams in the early 1970s, with Reed, DeBusschere, Bradley, Frazier and Barnett. Our dad commuted to work in New York City, and occasionally would bring home a Knick program from Madison Square Garden that he would buy from the vendors, much to our delight. We made sure to visit the Garden for an occasional Knick game. At the same time, the Nets were playing in the ABA league on Long Island. Although I never met any of the Knick players in person, I did get to meet several NY Net basketballers, including Ollie Taylor, Rick Barry, Jim Ard, and Joe De Pre during their summer promotional camps.. Later, while a graduate student, Dr. J (Julius Erving) traveled to Virginia Tech and I shook his hand and spoke to him – a highlight of my young life, in addition to playing in a pick-up game with Ollie Taylor (Nets) and Tim Hurtin at Newbridge Road Park on Long Island one summer evening. It was a great time for basketball, and both my brother and I enjoyed playing, and possessed above average jumping ability – probably due to having a greater number of fast twitch muscle fibers in our calves.

Aside from sports athletes, I was aware of bodybuilders and wrestlers. I recall Freddie Blassie, The Grand Wizard, Gorilla Monsoon, Bobo Brazil, Hans Schmidt, Beautiful Bobby, Lou Albano, Superstar Billy Graham, Ernie Ladd, and of course Bruno Sammartino. Then, there was the impact of Hollywood. Specifically, the Hercules movies of the late 1950s, starring bodybuilder Steve Reeves. I imagine every boy who saw these movies wished he could be like Mr. Reeves. I also had a healthy admiration and respect for the actor who portrayed the Man of Steel on television, George Reeves – even writing "The George Reeves Adventures of Superman Companion" book. Let me not forget muscly actor William Smith from the era of the 60s-80s, who appeared on television in "Laredo" as Joe Riley and in a multitude of TV series and movies ("The Losers", "Piranha, Piranha", "C.C. & Company") and even a Clint Eastwood film ("Any Which Way You Can").

I stopped playing recreational basketball when I was in my early twenties when life got serious. And following a period of searching for meaning in life, I met the woman of my dreams and we both strongly embraced married life and Christianity. We ended up attending Virginia Tech and graduating at the level of Ph.D., and I earned the doctorate in Human Nutrition and Foods in 1989. After a five-year work stint at Iowa State University, my bride and I traveled south to Texas A&M University in 1995, where we still teach and do research as professors. During a three and one-half year hiatus period from the university (2001-2004) when former Secretary of Defense Robert Gates was A&M president, we were employed at the federal level working at the USDA in the George W. Bush administration.

During the decades that followed my college days, I maintained a rather sedentary lifestyle. I did not strength train with weights until I reached 63 years of age in 2017. But I am so glad that I eventually did, because doing so has changed my physique as well as my physical and mental health. As the expression goes, "better late than never"! Specifically, I knew that I needed to do something to get into shape when in 2017, I was invited to host a nostalgia quiz in Sarasota, Florida. This was for a wonderful group of TV Superman (George Reeves) baby boomer enthusiasts, led by Mr. Jim Nolt ("The Adventures Continue"), and with the able assistance of Mr. Larry Ward and Mr. Brad Shey. The reason I had the awakening regarding getting in shape had to do with viewing the video footage of the event several weeks later. Sure, I already knew I was overweight and actually obese at close to 250 pounds, but seeing myself in the video really jolted me - what a gut. And skinny arms to boot.

At first, I was unsure of just what to do. I knew professionally, and could advise others, regarding following a healthy eating plan and incorporating exercise. But when it came to adopting such a lifestyle myself, I had been a failure. It was indeed time for a change. But what? How? I did not want to join a gym – for a variety of reasons. First, everyone at those places seems to be young and already fit, and I knew that would make me feel old and hopelessly out of shape! This would not be good for my self-esteem and motivation. Also,

I knew that I needed an age-appropriate training program, one that would not cause me to become injured and discouraged, eventually giving up all together. I also had a feeling that "old school" methods worked – recalling how Steve Reeves and others of his era trained very simply, yet with great results. I just wanted to find a way that captured that essence for me to incorporate into my lifestyle as I grew old. Not to attempt to become a bodybuilder at my advanced age, but rather to reap the known and proven health benefits of strength training, as incorporated into what I have branded my Old Man Muscle™ strength, fitness and nutrition workout routine.

Why focus on building strength? The answer has much to do with the multitude of health benefits, including those to the cardiovascular system, the musculoskeletal system, the brain, plus improved balance and flexibility that come with strength training as it does with attempting to add strength and muscle. While we are all aware of the concept of being "healthy", we also should realize that there is such a thing as "metabolic health" to consider, and that strength training can help achieve it. We are metabolically healthy when we are not needing any prescription medications, plus have optimal levels of triglycerides, blood glucose, blood pressure, high-density lipoprotein cholesterol (HDL), and possess an acceptable waist circumference. These are all potential risk factors for poor cardiovascular health. In other words, to be in good metabolic health means that our heart and blood vessel health is optimal, and in addition that we are not diagnosed with diabetes. A last point is that we should not confuse fitness with health. A person might be able to achieve a reasonable level of physical fitness, but if they suffer from cancer, diabetes, or hypertension, they are unhealthy.

As many readers are aware, there are a zillion books out there, in addition to videos and articles and blogs to read on the internet, regarding exercise. Everyone has an opinion. And, there are so, so many different exercises to choose from. What I ended up doing was gathering as much information as I could, looking always to see if there was a scientific evidence base for it. I have learned quite a bit regarding exercise from the graduate students and professional colleagues in the Health and Kinesiology Department at Texas A&M University that I interact with. I gradually distilled it all down and identified what impressed me as the foundation exercises that would enable me to strength train safely and effectively as an untrained novice in his sixth decade of life. A key aspect of the workout was that it needed to be brief – not taking 60, 40, or even 30 minutes.

The result was a ten-minute workout, three days per week that never left me bored, and that I could fit into my schedule. Not only did I lose the weight mentioned, but I added muscle. And I looked forward to the workouts, and was upset whenever I had to miss one! As I altered my body composition away from fat and toward muscle, I slowly but steadily began reshaping my physique. My blood pressure dropped from 145/90 to 107/70, to my great delight. I felt more energetic, and with better clarity of thought. I felt more like I did when I was a younger person.

So that is what this book is all about – sharing my knowledge and the training technique that I developed based upon my own reading of the research, my understanding of the old but proven techniques that work, and my application and testing of them in my own life. I realized that my simple ten-minute workout needed to be shared with other old men who, like myself, were heading down the wrong path that would eventually lead to physical weakness, frailty, and a low quality of life. I have included other information about aging and a unique section on spiritual strength training that complements the strength training information and contributes to overall health development and cultivating a positive outlook – in spite of being "old". It gives me great joy and pleasure to do so within the pages of this little book, and I trust the information will be well received and of value to you.

So – to my fellow "old men" out there – as we face the challenges of aging, I hope that you can make good use of the information I will be providing, and that it changes and improves your overall health, strength and physique, your mental and even spiritual outlook, and your enjoyment of life. Speaking of aging, the next chapter considers in some detail the challenges that arise as we become older men. It also summarizes some of the benefits of strength training in meeting these aging challenges, so please join me as we continue our Old Man Muscle adventure together.

RECAP AND WHAT'S AHEAD

I have provided you a little background on who I am and how I happened to put this book together. I also shared with you my hope that any reader interested in improving their health status should consider strength training. The Old Man Muscle workout, the core feature of this handbook, has worked for me, and I hope it can work for you too. In the next chapter, I will present information on the challenges that each of us face as we age, and the consequences. I will prepare the reader to expect that strength training can become the cornerstone of a lifestyle strategy that will enable us to meet many of the challenges of aging head-on.

THE CHALLENGES WE FACE AS WE AGE

In this chapter, the emphasis will be on the negatives that accompany aging. I will be calling them challenges, or aging consequences. But before I get into that material, I want to introduce a solution to many of these aging issues, and that is strength training. Readers take note: we will have much more to say about this form of exercise in the next two chapters. But I wanted to open our minds right away to this topic, which is why I have placed it here.

If you are reading this book, it is likely that you are either a baby boomer or a member of an adjacent generation before or after. So, you are reading about aging consequences that you have already experienced or may experience one day in the future. Now, it is interesting that many of us have noticed how adolescents and those in their young adult years have recently flocked to physical training. This has created in our generations a similar desire, especially when we realize that we struggle with health conditions such as diabetes, obesity, hypertension, arthritis and so on. These can be addressed through physical fitness and an active and properly nourished lifestyle. In this chapter we will briefly examine some of the primary reasons to make strength training a priority in our lives. We will also consider some key benefits that can be gained when we commit to this form of exercise.

Generation Categories in the USA	
Generation X	Born between 1965-1980
Approximate age range of adults in 2019	39-54
Baby Boomer Generation	Born between 1946-1964
Approximate age range of adults in 2019	55-73
Silent Generation	Born between 1928-1945
Approximate age range of adults in 2019	74-91

▲ **Table 2.1** Chart showing three adjacent generations of likely readers of this handbook.

As we continue, let's understand what we are getting into – and why. What is meant by strength training – what is it, and what is it not?

WHAT IS STRENGTH TRAINING?

Strength training, also known as resistance training, refers to training the body in such a way so as to increase skeletal muscle strength. This requires that the muscles work against and overcome resistance loads, such as weights. We lift them in opposition to the downward force that gravity places upon them. Although many readers may immediately equate this to bodybuilding, it is not exactly the same thing. In strength training, especially for older men, the goal is to become physically stronger in order to maintain functional movement and make daily activities easier to manage. Bodybuilding on the other hand is a sport (typically for younger men) in which the person trains to develop the physique to maximal levels. While it is true that when training as a bodybuilder to produce muscle growth there will be strength gains, it is not always the case that bodybuilders of a certain height and weight are as strong as weight lifters (powerlifters) of the same height and weight, due to differences in training. Furthermore, strength gains come not just from muscle growth, but from training the nervous system to make it more efficient and effective when it comes to sending signals to muscle tissue.

Strength training can be implemented through a variety of methods (often called "modalities"). These include the use of free weights (barbells, dumbbells, kettlebells, medicine balls), machines that utilize weight stacks, cables, pulleys or bands, and through body-weight or immovable resistance (isometric) exercises. The common feature is that when we work out, we consciously and purposely apply force to overcome a resistance load. All of the modalities mentioned except for isometrics involve motion – the movement of our joints, which in conjunction with bone and muscle, act as levers. The force that generates movement against the resistance load is of course coming from our muscles, which themselves only move because of the action of neurons (the nervous system) upon them.

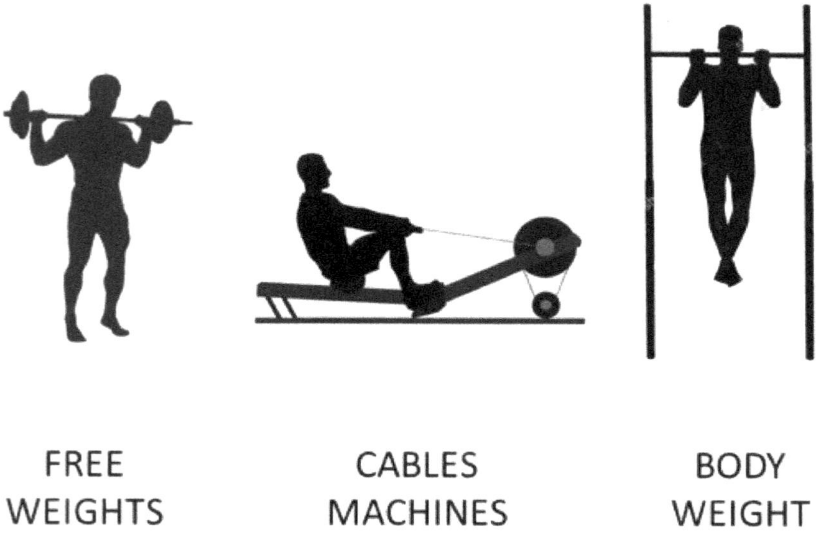

▲ **Figure 2.1** Three popular training modalities, using either free weights, equipment of some kind (cables, pulleys, weight stacks, bands), or body weight.

All of these strength training approaches create a challenge and a stress upon various systems of the body, including the cardiorespiratory, endocrine, muscular, nervous, and so forth. But, unlike the majority of the challenges that will be described in this chapter, this one is good, and beneficial. An exercising person transitions from a state of rest to the workout, and specific strength training exercises result in body adaptations that are desirable. During the Old Man Muscle workout, the heart will pump harder and faster to meet the demands of the exercises. And so there is accelerated blood flow to deliver oxygen and nutrients to exercising muscle, which provides the stimulus for exercise performance. There can be an increase not only in blood flow, but also temporarily in heart size and blood pressure, all of which are normal and healthy situations. However, because the exercised state is a stressed one compared to the pre-exercise resting state, it is always necessary to warm up the cardiorespiratory system, as well as the joints and muscles, prior to any effort against resistance loads in order to prepare them for change and to avoid any adverse consequences. You will notice when the Old Man Muscle workout is explained in Chapter 4, that we do incorporate a warm up routine plus a pre-workout stretch in our workout as detailed in the next chapter, and it is precisely for this reason.

To recap, in our little discussion regarding strength training so far, we equate it to a form of exercise (or "working out") that is called resistance training. This is because in order to increase strength, the muscles of the body will be worked in opposition to a resistance load, such as dumbbells (as well as working in opposition to gravity), which will generate strength gains.

Some readers are immediately wondering if strength gains is the same thing as muscle size (**hypertrophy**) gains. Although related in terms of being aspects of muscular fitness, muscular strength and size depend on slightly different training protocols for enhancement. There is also the aspect of muscular endurance, and its specific training protocols and effects. As previously implied, large-muscled bodybuilders can be less strong than smaller-muscled Olympic weightlifters or strength athletes, due to differences in muscle physiology and training programs. One person trains for size while the other trains for strength. Having said all of this, the Old Man Muscle workout, like many others, allows for both strength development and hypertrophy. So, it is fine if the reader wants to think of himself as participating in the sport of bodybuilding, especially if that motivates him to continue to lift and become stronger in order to sustain functional independence in old age.

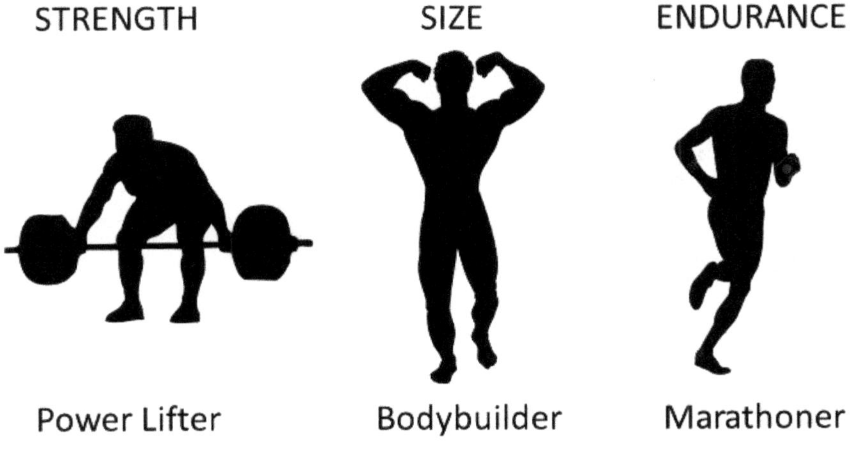

▲ **Figure 2.2** Three aspects of muscular fitness: muscular strength, muscular size, muscular endurance, and representative body types. Note – although it is possible to train for all three of these depending upon the training protocol used, most individuals select only one or at most two of these as goals, and train accordingly.

Consider also, that the endurance athlete (like a competitive marathon runner or cyclist) has a distinctive muscle training approach that specifically develops aerobic (cardio) fitness. Here, there will be less muscle size and strength overall compared to body builders and power lifters. Understanding the difference enables proper program design for the intended fitness outcome. As indicated, the Old Man Muscle exercises and training protocol is something of a blend of muscle training approaches, and can provide for both increases in strength and moderate muscle hypertrophy, and even some aerobic endurance if the routine is completed at a sufficiently fast pace. Although performing endurance exercise, even in old age, can have benefits, research has demonstrated that strength training rather than cardio should be emphasized in order to combat frailty and the risk of injury. There are unique benefits to strength training compared to cardio, which makes strength training our best priority for exercise, with cardio still important but to a lesser degree.

THE CHALLENGES OF AGING

Next, we will discuss the reality of the aging phenomenon that we are currently experiencing. As we all know, aging is a natural process of life. Our aging body - that trusted and reliable companion for so many years, changes and undergoes a process of decline as we grow older. Aging, one might say, is programmed into our DNA. It is non-negotiable, and ultimately this leads to an expiration date to life that we all will experience. At the later stages of aging, we begin to become very aware of the body's weaknesses and limitations, and this requires an adjustment on our part, both in thinking and in terms of action.

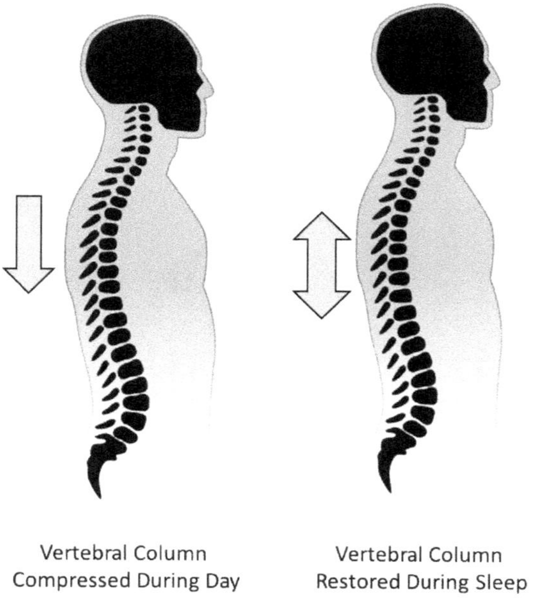

▲ **Figure 2.3** A height differential exists depending upon the time of the day, due to compression of the spine caused by the force of gravity and posture. Height diminishes somewhat as the day progresses due to the body being upright, but is restored after a good night's sleep due to being horizontal. Don't sleep sitting or standing up!

Aging (also spelled ageing - the scientific term is senescence) represents the sum total of processes that begin to decline basically once we reach adulthood. These processes are biological and physiological (the functioning of body cells, tissues, organs); physical attributes and functioning (height, weight, strength, balance, mobility, flexibility, appearance); mental and psychological functioning (cognition, mood) and more. Aging can also be viewed as the accumulation of damage in molecules, cells and tissues over a lifetime that compromises functional ability and increases risk for many diseases and for death. More simply stated, aging is a time-related deterioration of what we were as a young person to what we become in our later years.

Since I mentioned height, allow me to make a point about height loss. Most of us realize that we lose height as we age – but did you know about the daily fluctuations? Yes, fluctuations in body height occur throughout the day, and they are due to posture and flexibility factors. It is known that the vertebral column shortens throughout the day perhaps as much as half an inch, and regains its length when the body rests/sleeps lying down at night. The change is due to movement of water and associated disk fluid in and out of the intervertebral disks of the spine. During the day, due to our upright posture, the force of gravity, and demands placed upon the musculoskeletal system, disk dehydration occurs as fluid is squeezed out and the spine compresses.

On the other hand, during sleep, the horizontal spine decompresses and the disks become rehydrated and swell with fluid, restoring our full stature upon arising in the morning. Due to disk swelling, our spine (particularly the lower, lumbar region) can feel stiff upon awakening. As we become upright and begin to move, we experience less stiffness and greater mobility throughout the day. So working out first thing in the morning may feel different compared to working out when we have greater flexibility and mobility as the day progresses. Risk of injury to the intervertebral disks may be increased in old age if we try to lift weights right after we arise – especially if we fail to warm up and perform some flexibility exercises first.

Alright; let's get back to the challenges of aging.

Fortunately (or unfortunately), aging begins more or less as a gradual decline. In some, it is more gradual than in others. It can sneak up on you. But it will happen to everyone, and accelerates the older we become. The question is, what – if anything, can be done about it? The quick answer is that we can do nothing about the ultimate outcome – but we can impact the journey through our final years that brings us there, as long as we have a sound strategy.

Obesity	Decreased vitality
Diabetes	Anabolic resistance
Sarcopenia	Poor concentration
Hypertension	Vitamin D deficiency
Osteoporosis	Decreased sexual drive
Low testosterone	Increased risk of falling/frailty

▲ **Table 2.2** Some of the major challenges that we face as old men.

Before I continue, I thought it would be nice to inject some humor about being old that I have either been told or otherwise gathered from various sources. For instance, I've noticed – and perhaps you have too - that the older we get, the earlier it gets late at night! And when I looked in the mirror this morning, I thought that what I saw was just a result of being really tired today. But since it's been the same look for the past five years, I guess this is the way I really do look now!

I can remember back to when I would refer to my knees as "right and left" instead of "the good one and the bad one"! And you know you're getting old when you are exhausted from barely doing anything all day, and need to take a nap just so that you can wake up later with enough energy to continue to do essentially nothing! Men, let's come clean. You know you are old when:

- *The gleam in your eyes is from the sun hitting your bifocals.*
- *You start discussing your bowel movements with your spouse.*
- *Your little black book contains only names that end in M.D.*
- *It is Friday night and you look forward to a dull evening at home.*
- *You are constantly discussing the price of gasoline around town.*
- *Your ears are hairier than your head.*
- *And... the little old lady you help across the street is your wife.*

Well gentlemen, we've had some fun, and I hope some of those statements gave you a little chuckle. Time now to ask you a simple series of more serious questions:

(1) How old are you? (your chronological age in years)

(2) Are you at a healthy weight?

(3) How sharp is your mind, and your memory?

(4) Is it easy to get up from a sitting position, and are you flexible?

(5) How would you rate your upper body, core, and leg strength?

(6) How old do you look?

(7) How old do you feel?

Okay, so these are some tough questions that quickly point out to all of us that we are not getting any younger, and that we have issues now that we did not have to face when we were in our twenties. Notice - I did not even ask how many medications and pills you are taking!

From these simple questions, we move to another: are you out of shape? What even does being out of shape mean? On one level, it can mean not being in the same shape today that you were when younger. More specifically, it can mean that you are in poor physical condition, and not physically fit. I remember as a kid (back in the days of President John F. Kennedy) that in school we learned about the President's Council on Physical Fitness. I distinctly felt a pressure that the President of the United States would somehow find out if I was in good shape or not – and feared that I would fail his test! This same entity is now called The President's Council on Fitness, Sports and Nutrition. The goal of the PCFSN is to promote, encourage and motivate Americans of all ages to become physically active and participate in sports. Well yes – for us, as men of advanced age, we can fulfill this mission because our sport will be strength training, through the Old Man Muscle workout.

You Might Be OUT OF SHAPE if:	
You feel tired all the time.	You cannot touch your toes.
You get winded climbing stairs.	You cannot see your toes.
You can't do push-ups- or chin-ups	The thought of exercise makes you wince.
You have abig waistline.	You do not have pain-free movement in joints.
You ask people for assistance.	Upon exertion, it takes a long time to catch your breath.

▲ **Table 2.3** Are you out of shape? Here's how you can tell.

We see that aging is a time-dependent progressive and degenerative process. It has been proposed to be due to a multitude of factors, including inflammation, oxidative stress, free radicals, and **telomere shortening**. However, cause and effect are as yet not crystal clear, and no single theory fully explains human aging.

Genetics

The reader might immediately wonder about the role of heredity (genetics) in all of this. After all, we inherit so many traits from our parents due to the genes we possess. And, due to each person's unique genetics as well as habits, not everyone ages at the same rate. Someone who is in their forties chronologically may have aged more than another person who is in their fifties and appears more youthful.

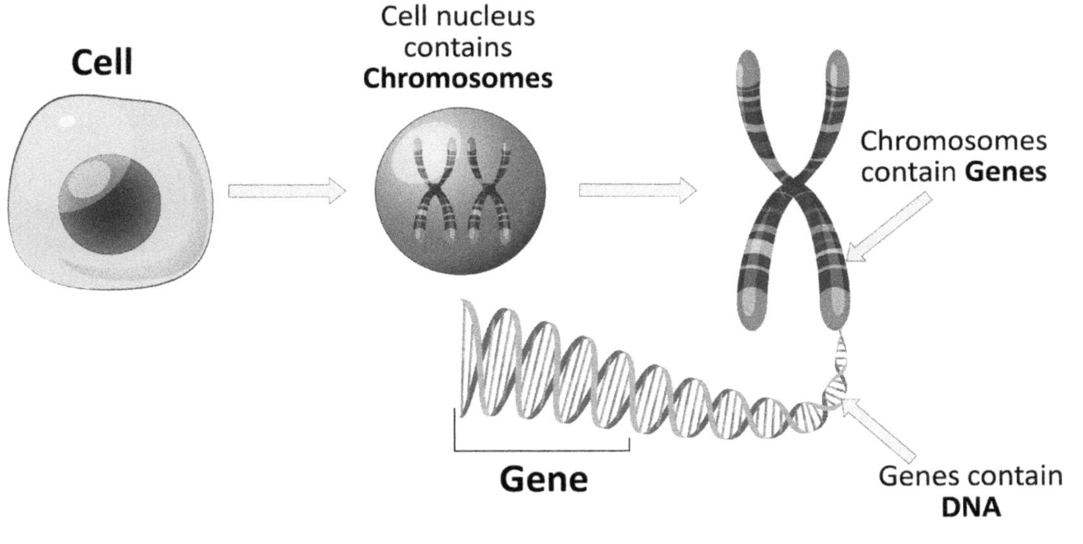

▲ **Figure 2.4** The chromosomes contain our hereditary material as DNA within genes.

On the point of genetics and heredity, I simply must take a moment to illustrate the concept by making a reference to one of my favorite television programs when I was growing up – The **Three Stooges**. There is an episode starring of course Moe, Larry, and Curly called "Hoi Polloi". Two high class gentlemen are having a discussion over dinner at a posh establishment. One asserts that environment is "the keynote for social distinction", while the other man insists that heredity is the reason. The environment man wagers the heredity man $10,000 that he can take three uneducated, lower class men – the Stooges - and transform them into refined gentlemen, regardless of their genetics. Well, I don't have to tell you how it all turns out! By the way – search for this episode on YouTube, it has been colorized, and is a hoot. The point being, we owe a lot to heredity to be certain, and all else equal, it is tempting to think that the manner in which we age has a lot to do with the genes we inherit from our parents. However, genetics is not the entire story at all, as the environment argument would tell us (even if, in the case of the Stooges, this argument did not prevail).

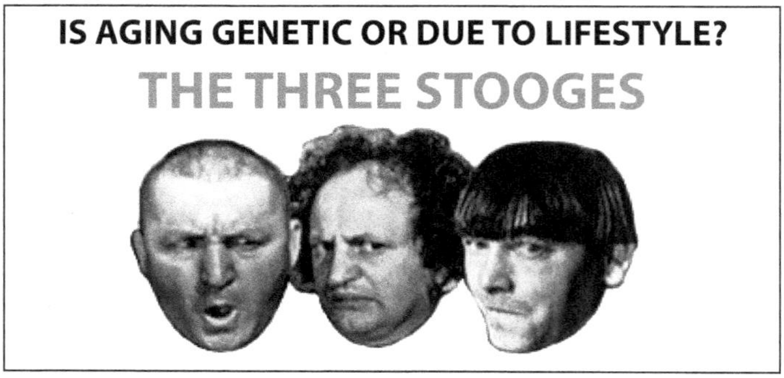

▲ **Figure 2.5** The Three Stooges raise a relevant question.

The current understanding of biological aging is that some aging-associated changes are indeed genetic (programmed), while others are unpredictable and influenced by environmental (epigenetic) factors. We will begin the discussion in a moment with epigenetics, and by the way it is believed that it may be possible to override some of our genetic programming by lifestyle choices (meaning, epigenetics - our diet, our physical activity, our rest and **sleep habits**, and so on). Thus, human senescence is indeed a complex process involving both environmental and genetic and influences affecting physiological pathways that impact how we age.

Now regarding the **epigenetics** of aging – which for our purposes describes the impact of the environment and lifestyle on body cell activity over time, particularly when it comes to genes being activated or deactivated. There are two specific cellular mechanisms: one is called gene methylation and the other, modification of histone proteins. I know that this is technical biology information and many readers may not be interested or have a biology background. However, I will try to be brief yet clear on this important topic so that we might better understand the aging process. As a reminder, genes are what contain the genetic, inherited material (DNA) within our body cells. Specifically, chromosomes contain all our genes, and genes are responsible for expressing (producing) proteins in that body that regulate how we look (phenotype), our physiology and metabolism traits, as well as potentially our personality and how we age.

The two kinds of epigenetic modifications mentioned above can occur at any time during our lifetime, and are responses to signals that our cells receive, such as stress as a result of our lifestyle behaviors (smoking, foods and beverages consumed, supplement use, hormones, and probably diseases). These can have a temporary or long-term impact on gene expression. **Epigenetic modifications** can take many forms and result from the work of specific enzymes that add small chemical groups like methyl groups to our DNA, either directly or through DNA-associated proteins called histones. The most-studied modifications are methylation (methyl CH_3 group addition), glycosylation (sugar molecule addition), and acetylation (acetyl group CH_3-C=O addition).

First, what is the meaning of histone modification? **Histones** are proteins around which the DNA associates within chromosomes. Modifications to histones (acetylation and methylation) can affect the body and change gene stability and expression. The modification of histone proteins changes the DNA:histone association such that DNA regions that were previously hidden or tightly wrapped by these proteins become exposed. The effects of these changes have to do with silencing (halting) gene expression.

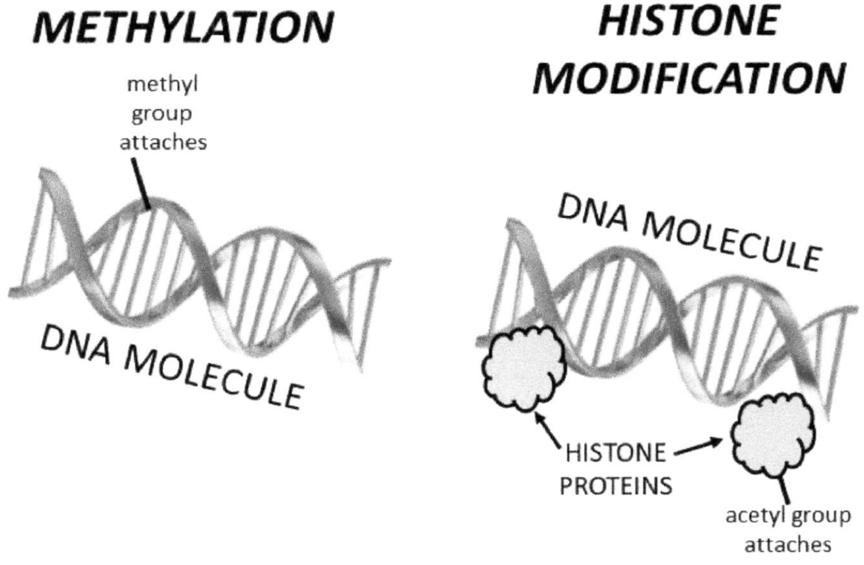

▲ **Figure 2.6** Two epigenetic factors thought to contribute to aging: DNA methylation and histone modification.

Second, gene methylation, also called **DNA methylation**, is the epigenetic phenomenon through which some genes can be modified with respect to methyl groups. DNA methylation is an enzymatic process that works directly on the DNA by adding CH_3, and this is powerful enough to cause the genes under effect to switch on or off. Said another way, the result of such a change to a cell's DNA can be to inhibit or silence gene expression.

So, we might well wonder, is DNA methylation a good or bad thing, and is histone modification a good or bad thing? These are great questions to ask.

Histone modifications can be considered "good" when the result is silencing of a gene that predisposes a person to disease, and "bad" when the reverse happens. Regarding methylations, since methylations perform critical functional roles, they might be assumed to be "good". During human growth and development, certain methylations are indeed essential in order to ensure a healthy developmental outcome for the individual. And although we inherit the extent of our DNA methylation, it can be altered throughout life by the environment and our lifestyle. And now, the idea of "bad" must be considered. It has been suggested that abnormal DNA methylations have adverse outcomes, as too much methylation (hypermethylation) of genes has been associated with cancer.

Specific to aging, the DNA methylation process happens more often as we get older, which inhibits gene transcription and expression. So as we increase in years, it may well be that hypermethylation is one of the causes of what we see as the aging process. This could mean that if hypermethylation can be delayed or altered due to some external factors of environment or lifestyle, we may be looking at an epigenetic therapy to manage aging. In all of this, it is tantalizing to consider that it may be possible to manipulate both DNA methylation and histone modification in order to create more favorable outcomes – such as to delay aging and maintain vigor and vitality longer. But much more research and study is needed before any solid conclusions can be drawn, or strategies applied in daily life regarding attempting to impact DNA through epigenetic means in order to enhance health.

Chronic Inflammation and Aging

We have all heard of the term inflammation, but what does it mean? **Inflammation** is the body's natural response to foreign substances (example: bacteria) as well as to cell and tissue damage. There are generally two inflammatory states, one which is brief and is termed acute inflammation, and the other which is long-term, called chronic inflammation. While acute inflammation is essential for healing and is considered "good", chronic, unresolved inflammation is believed to be associated with the development of age-related chronic disease, which makes it "bad". This type of inflammation may be caused by exposure to food allergens, bacteria and viruses, stress, and toxins.

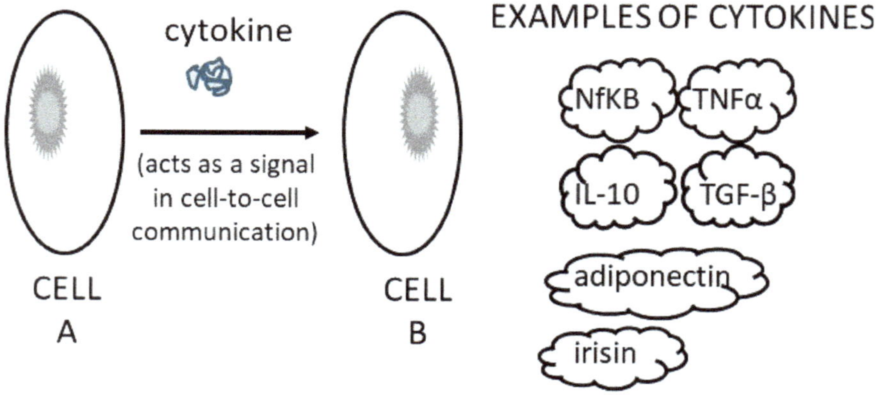

▲ **Figure 2.7** Cytokines are signaling molecules similar to hormones, that are produced by the cells of the immune system, as well as by a diversity of body organs, tissues, and cells, including muscle, fat (adipose tissue), lungs, liver, spleen, brain, and intestine. Shown are several examples. In inflammation there is an increase in circulating pro-inflammatory cytokines over anti-inflammatory types. Acute (short term) inflammation is beneficial, whereas chronic (long term) inflammation is not.

In attempting to understand inflammation, and for our purposes we will be superficial here, it is important to discuss the body's immune system. The immune system (consisting of the lymphatic circulation, lymph nodes, T lymphocytes, B lymphocytes, macrophages, the cytokines they produce, and the antigen:antibody response) represents the body's defense against disease and injury. But again – when the system acts quickly, this is healthy, but if the immune system is stimulated for prolonged periods, then we are dealing with chronic inflammation, which is unhealthy.

How aging comes into the picture is as follows. In older persons, there exists a chronic, low-grade inflammation. Scientists have coined the term "**inflammaging**" for this, which combines the words inflammation and aging. Inflammaging is a major risk factor for disease and death in the elderly population. In addition, it is believed that most if not all age-related diseases (cancer, diabetes, heart disease, obesity) are linked to chronic inflammation. However, at present unclear to what extent chronic inflammation influences the pace of aging.

Another term related to inflammation is **immunosenescence**. You may have guessed that this term refers to the fact that as we get older, our immune system also becomes old and shows signs of declining function. When this happens, the person is susceptible to infections and is at risk to become ill more frequently. I would like to mention here that proper nutrition is essential for optimal immune response and host defense against infection. Furthermore, it may be the case that certain exercise as well as nutritional interventions in old age may restore the functioning of an aging immune system. Stay tuned for more on the issue of boosting the immune system through diet and exercise in the next couple of chapters. Now - the next topic to be discussed, oxidative stress, is linked to inflammation and indicates that health declines due to aging are indeed complex.

Oxidative Stress

At some point in our schooling we all learned how important oxygen (O_2) is to life, and that we breathe it in through the air while we exhale carbon dioxide (CO_2). However, it turns out that oxygen is a harmful substance, metabolically speaking. Therefore, one theory as to why we age is that aging is due to the damaging effects of oxygen on our cells and key molecules that are linked to body metabolism. Metabolism, simply put, is the sum total of all the chemical reactions (enzymatic and nonenzymatic) that occur in the body on a daily basis. So according to the theory of oxidative damage, also called **oxidative stress**, it is oxygen that acts as a stressor, and aging is seen as a normal by-product of metabolism.

To get a better understanding, we have to appreciate what substances called free radicals are. Free radicals are chemical species formed during normal cellular functioning, and are highly reactive. A key example of free radical species are called **ROS** (reactive oxygen species), which includes the superoxide ion ($-O_2^{\cdot}$), the hydroxyl radical (-OH), and the peroxide radical ($-O_2^{\cdot}$), ROS are important because they actively oxidize and damage cell membranes, proteins (including enzymes), and nucleic acids (DNA). Due to this understanding, the involvement of ROS in aging has been proposed.

The body has a natural defense against the free radicals that cause oxidative stress, and this defense is the presence of endogenous (present in body cells) antioxidants. It is the job of these antioxidant molecules to neutralize free radicals. Many people purposely consume foods and supplements rich in antioxidants like vitamin C, vitamin E, plant (fruit, vegetable, grain, and spice) polyphenols, and even food additives like BHA and BHT. These are referred to as exogenous antioxidants because they are obtained from substances outside the body through the diet.

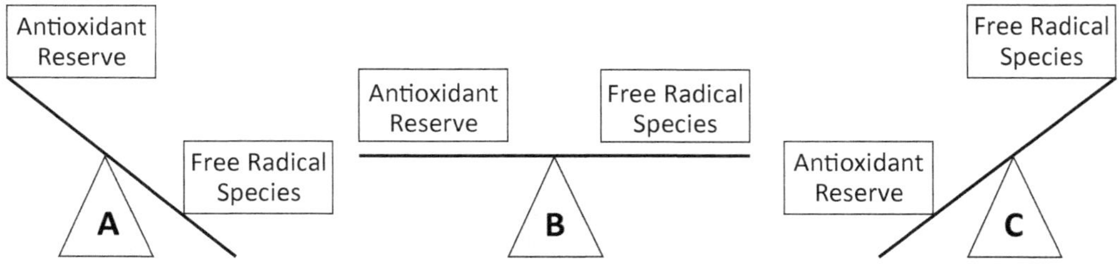

▲ **Figure 2.8** Three situations showing the body's antioxidant capacity to fight against free radicals and oxidative stress. In situation C, the antioxidant reserve is greater than the free radical threat, resulting in no net oxidative stress; in situation B, there is a balance between the body's antioxidant reserve and the free radical threat; in situation A, the amount of free radical species exceeds the body's antioxidant reserve, and oxidative stress occurs. Interestingly, even though exercise itself generates ROS as well as pro-inflammatory cytokines, the body becomes progressively more efficient at quenching these. This results in reduced baseline oxidative stress and inflammation in the trained individual.

When the sum of your endogenous + exogenous antioxidants is greater than the amount of free radicals being formed in your body, oxidative stress is nullified. However, if our antioxidative capacity ever becomes overwhelmed by too many free radicals, this is when oxidative stress – and the associated damage – occurs. Over time, the accumulation of such damage has been proposed to contribute to aging and age-related disease. This is the basis of what is called the **Free Radical Theory of Aging**. This theory has been long accepted since are numerous studies demonstrate that ROS and oxidative damage increase with age, and shorten lifespan.

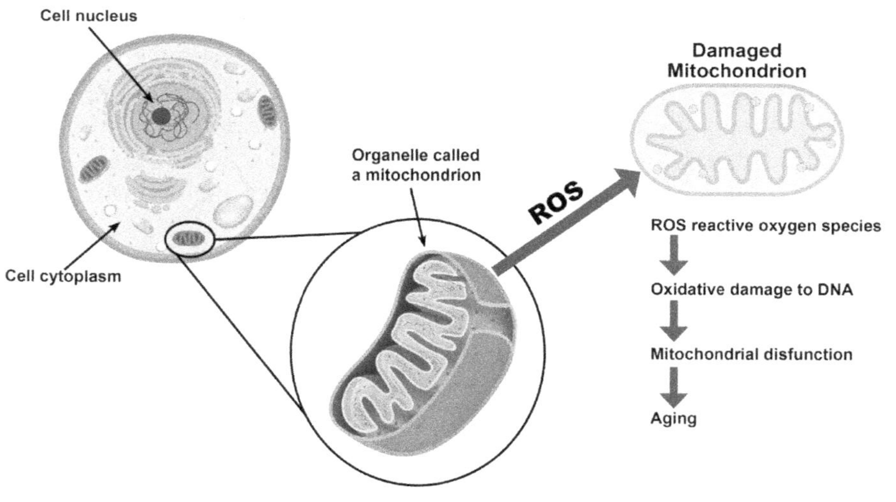

▲ **Figure 2.9** The mitochondria are "organelles" located within body cells out in the cytoplasm, not the nucleus. Mitochondria function to generate ATP energy as fuel for body cells, and also signal or otherwise direct other vital cellular components and processes. Cells that require the most ATP energy including skeletal muscle, the brain, eye and heart - contain the most mitochondria. These organelles have their own DNA, which can potentially be damaged by ROS. As it turns out, mitochondria are quite susceptible to free radical attack – which is basis of the MFRTA theory.

Interestingly, this theory has recently been called into question – in part due to a salamander. What? Yes – scientists have studied a species of European cave salamander that has exceptional longevity, yet lacks robust defense against oxidative damage. In addition, it has been suggested that other types of non-oxygen related damage that contribute to "biological imperfectness" may be important in aging. In this view, cellular components and systems are viewed as imperfect and heterogeneous, and capable of generating variable end products ("metabolites") routinely as part of normal metabolism. The accumulation of such imperfect metabolites over time leads to deviations from what is "normal" and "healthy" in cellular processes, which, over time, contribute to aging.

There is an offshoot of the Free Radical Theory of Aging that I wish to mention briefly, in case you ever hear of it beyond the pages of this book. It is the **Mitochondrial Free Radical Theory of Aging** (MFRTA). It states that free radicals generated in body cells within the mitochondria as normal by-products of metabolism cause oxidative damage. Such free radicals are termed mitochondrial reactive oxygen species (mtROS). Further, that the accumulation of this kind of oxidative damage is behind the aging process. To be clear, this theory is not universally accepted, in part due to the fact that experimental animal studies have failed to demonstrate the expected extension of lifespan in animals that generate fewer mtROS in their lifetime.

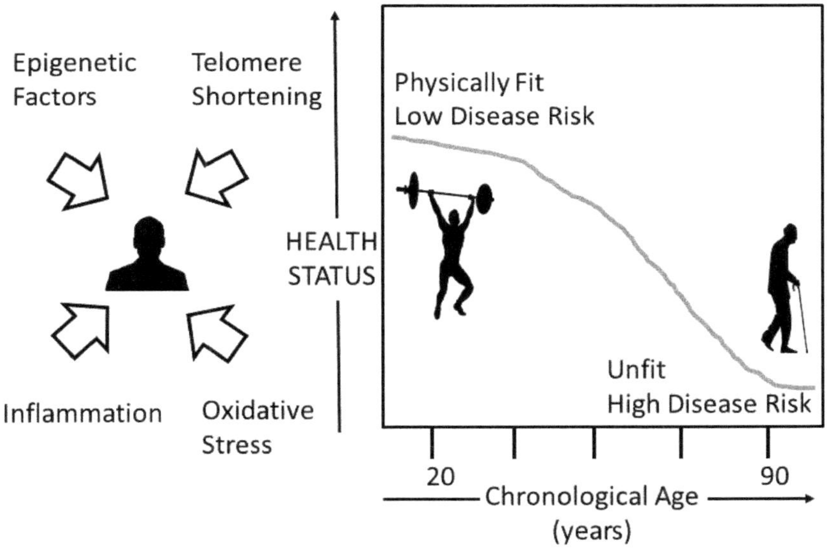

▲ **Figure 2.10** Factors that affect aging and the associations between aging and fitness/health. All theories of aging recognize an imbalance between stressors (epigenetic mechanisms, inflammation, oxidative stress) and the body's compensatory actions, which decline over time leaving unrepaired damage at the molecular and cellular levels and beyond. The end result is decreased disease and stress resistance, gradual physical and cognitive health decline, and finally failure to thrive. Shown is a hypothetical curve representing these changes over time.

Before leaving the topic of free radicals, oxidative stress and aging, I hasten to point out that exercise – especially when intense - has been identified as a source of oxidative stress. As such, it has been suggested that intense exercise, whether endurance (cardio) or to a lesser extent, weightlifting - increases the risk of heart disease, cancer, causes us to age more quickly, and possibly to die sooner. However, things are not as simple as they appear in all of this. It turns out that free radicals not only function to damage cells, but they act as signaling molecules that result in positive changes. What paradoxically their signals have been shown to cause is an increase in the body's antioxidant defenses over time, actually making the body more resistant to oxidative stress and damage.

So from the exercise perspective, a trained person who exercises with intensity has certainly induced a healthy amount of free radical production – but the key is, the body will adapt to this situation to become more efficient and effective in dealing with the oxidative stress compared to someone who is not trained and exercising similarly. When viewed in this manner, exercise-induced oxidative stress in a trained person is a desired consequence of the workout, because adaptation occurs to increase the body's antioxidant levels and routinely provide enhanced protection from oxidative damage compared to a person who is out of shape and lacks the adaptation.

Telomere Shortening

The only other proposed cause of aging that we will discuss is an intriguing one, and to understand it we need to look at DNA. Recall that DNA is the genetic material that makes us who we are, and is found in our body cells within the genes on chromosomes. We must also realize that our cells have a life cycle. Periodically, our cells will divide and form new copies of themselves – replication this is called. They must also make copies of their DNA when this happens, to put into the chromosomes of the next generation of cells. It turns out that there are regions of chromosomes called **telomeres** that researchers have learned become progressively shorter or "used up" through each successive cycle of cell division. Although the telomeres are part of DNA and chromosomes, they themselves do not contain any genetic material.

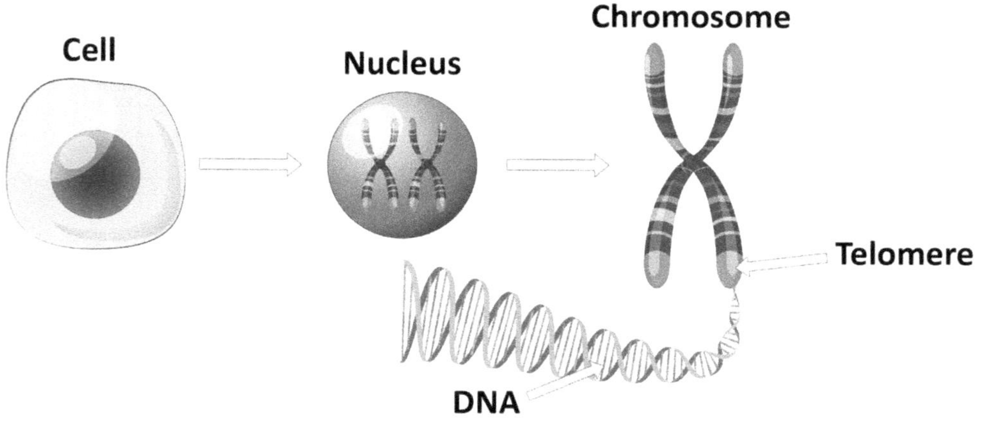

▲ **Figure 2.11** Drawing of a cell showing that within the cell nucleus are chromosomes made up of our DNA, and that telomeres are located at the ends of the chromosome structures. Telomeres are the end of each strand of DNA, and form "cap" regions on chromosomes to the DNA. Over time, they erode (shorten), which hastens the senescence and death of the cell.

When telomeres shorten to the extent that they can no longer function to protect the DNA in chromosomes, our cells age and they cease to divide. As such, telomere shortening has been proposed to be related to aging, with telomeres being considered to represent the aging clock in every cell. Telomeres shorten as we grow older, but research indicates that telomere shortening is impacted by lifestyle – as stress, lack of exercise and poor nutrition can adversely affect telomeres.

Furthermore, telomere length is an index of our biological age rather than our chronological age. If you were able to look at two cells, and one of them shows much shorter telomeres, then that cell is going to become senescent sooner than the other one. Now, is it reasonable to suspect that cellular aging of this type can be translated into overall aging - I mean, someone's visual appearance and susceptibility to disease and such? Recall that some people do not appear as you would expect, considering their chronological age. For instance, a certain person who is in their seventies may look younger than someone else who is fifty-five. Could this be due to the fact that their telomeres have resisted shortening (or may even have lengthened) due to healthy lifestyle habits compared to the person who looks their age and is overweight, smokes, eats junk food and has a lot of stress in their life?

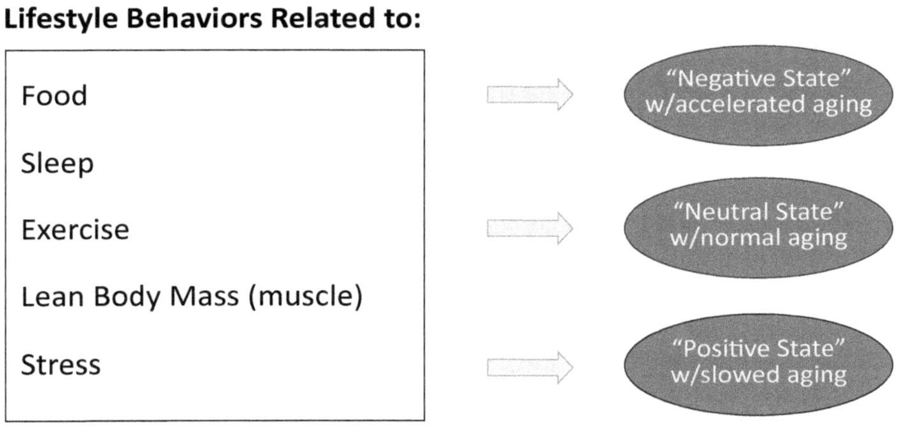

▲ **Figure 2.12** Five lifestyle behaviors - what we eat, how much we sleep, our exercise habits, the amount of lean body mass we have, and the amount of stress in our life – suggested here to be related to telomere length and aging. The "negative state" would equate to more rapidly shortened telomeres and accelerated aging due to poor lifestyle behaviors, while the "positive state" would result in decreased shortening of telomeres and delayed aging based upon favorable lifestyle behaviors. Normal telomere shortening and aging – neither shortened nor lengthened – would be the intermediate outcome in the "neutral state".

The phenomenon of telomere shortening, then, is believed to be a key marker of cellular aging – one that is potentially reversible. But is it simply a marker of aging, a consequence of aging, or a cause of aging? The answer is open to debate at present. However, my default decision regarding presenting it in this book is to consider it as a cause of aging, because when telomeres are not allowed to shorten (by this I mean when they lengthen instead), aging is delayed.

Nobel prize-winning scientist Elizabeth Blackburn (awarded the prize for her research on telomeres) has written a layman's book, "The Telomere Effect." The message of the book is "you have more control over your own aging than you may imagine." She believes this to be true and that a person can actually lengthen their telomeres — and perhaps their life as a result — by living a healthy life. It is known that an enzyme called telomerase is present in certain cells, and that this enzyme, when activated, causes telomeres to lengthen. So it is tempting to suggest that a healthy lifestyle somehow gets more telomerase involved to do its thing. In spite of this, there is less than universal agreement among scientists regarding the value of trying to lengthen one's telomere length, in part because the manner in which telomeres are measured in various studies has some problems associated with it. As a result, some scientists feel that it is too early to say if telomere lengthening and associated predicted benefits are "real".

As we close out this topic we also need to apply common sense and keep in mind that the human body is not an immortal machine, and that there are limits. Too much intense exercise generating too much oxidative stress can exceed what the human body can handle, even at the level of the elite athlete. But for many of us, we can train in such a way to maintain moderate to high intensity workouts that are age-appropriate, while following a good diet and allowing for ample rest and recovery periods, resulting in desirable adaptations that provide protection against potentially dangerous levels of oxidative stress.

AGING CONSEQUENCES

In this chapter, I have been describing the challenges of aging. We can think of them as aging consequences too, which are many and varied, as we have seen. Look again at that list of diseases and challenges, and potential physical limitations (Tables 2.2 and 2.3). A rather gloomy picture – but take heart, all is not lost. On a positive note, environmental and lifestyle factors (related to what is called epigenetics) have emerged as one area that is potentially under the control of the individual to help slow down the aging process, with physical exercise being a key strategy due to the desirable anti-aging adaptations it can cause. So, I am proposing that we view exercise (strength training in our case) as medicine – in that it can address aging to a certain extent to generate a more youthful and active aging experience, and can fix or prevent some age-associated diseases.

We have just reviewed some rather technical details regarding the causes of aging. What we will turn to next in order to finish out this chapter involves the consequences of aging, including anabolic resistance, changes in body composition and shape, sarcopenia with associated strength decline, testosterone decline, and vitamin D decline.

But before we discuss these, let's take a break and think back to what it meant to be young. It's okay to think about your friends and family here, as well as your school days, and about doing crazy things as a kid. However, I'd like to shift the focus once you are ready, to the matter at hand – how the younger body performed. It seemed invincible, right? So now, imagine we are back in our teens. You know that is considered the peak time of development, during which we have the advantage of a quick and responsive nervous system, a more supple and powerful musculoskeletal system, fantastic levels of growth hormone and testosterone, an excellent cardiovascular system, as well as joints, tendons, ligaments that are in prime functioning condition. All of this in our younger days gave us the best shot at obtaining results from strength training. This means everything from muscular development, enhanced strength, quicker and more powerful muscular response, better deep sleep, and all-around enhanced body functioning as well as recovery from all the stress of intense exercise and life in general.

In a nutshell, in those days we could run faster, jump higher, and last longer. But thanks to aging, we suffer losses in these areas. What are called neural drive, elastic energy, and muscle power output all become less as we enter our 30s and beyond. We just cannot maintain the physique, performance level, and stamina of our younger days. If you have ever watched the aging of an elite athlete - be it a basketball player, baseball player, or football player – the common denominator is, when they end up participating in their sport's "Old Timer's Game", they are a mere shadow of their former selves. And these people are trained athletes who were better than the best. It is a sobering thought.

Importantly, we need to realize that no strength training program will ever turn back the clock to recapture those days when we were 18 or 20 years old. It simply is not possible, no matter how much we train or how much we want it. So let's see what is going on regarding the known consequences of growing old.

Anabolic Resistance

We start off discussing anabolic resistance. What is it? To understand anabolic resistance, we must know what the term "anabolic" means, and we must also decide what role the word "resistance" plays here. Let's start with the word anabolic. Some readers will have heard about "anabolic steroids" and that people have used these drugs in order to build big muscles quickly. Well, in order to understand anabolic, we must briefly discuss metabolism. **Metabolism** refers to the chemical reactions that take place in the body that provide for digestion of food, absorption of nutrients, energy production, body growth and development, reproduction, and ultimately, good health and survival.

Metabolism can be subdivided into two phases: anabolic metabolism and catabolic metabolism. **Anabolic metabolism** refers to the phase in which molecules required by the body are built, or synthesized. This form of metabolism typically requires energy input (ATP energy that foods provide). On the other hand, **catabolic metabolism** refers to the phase in which complex molecules like starches, proteins, and triglycerides are broken down (hydrolyzed) into smaller molecules; this releases energy for the body to use, which can drive anabolic reactions. So now we have the idea that anabolic means building up molecules, while catabolic means breaking apart molecules. This thinking also transfers to tissues like muscle tissue: when we lose muscle mass, it is called muscle catabolism, and when we gain muscle mass, we are talking about muscle anabolism. On the whole body level, there are hormones that aid muscle growth – such as testosterone, and these are called anabolic hormones. There are also hormones that inhibit muscle growth, like cortisol, and these are called catabolic hormones.

Furthermore, many readers are realizing that when we exercise, we might be performing anabolic exercise when we lift weights, and catabolic exercise when we do cardio (endurance exercises). While catabolic exercises are considered aerobic because they require oxygen, anabolic exercises are considered **anaerobic**, because they do not require oxygen. From this, we see that a workout can have anabolic or catabolic features. When the intensity is high and the timeframe of the workout is relatively brief (say 30 minutes or less), then muscle mass will increase post-workout and fat will be burned – not only during the exercise, but also over the next day. These are the features of anabolic strength training. On the other hand, the features of a catabolic workout are when the intensity is moderate and the timeframe of the workout is relatively long (say 45-60 minutes or more). In this kind of workout the muscle mass is not increased as in an anabolic workout, while fat will be burned only during the timeframe of the exercise. Total loss of body mass (muscle + fat) is a potential feature of this kind of workout.

Actually, the two forms of workout complement each other and can be used together. A type of training called **HIIT high-intensity interval training** and also circuit training combines catabolic and anabolic exercises. For our purposes, the Old Man Muscle workout represents primarily an anabolic workout.

Well, now is time to define **anabolic resistance**. There are actually at least two types of anabolic resistance. One is the anabolic resistance to **MPS** (muscle protein synthesis), while the other is the anabolic resistance to exercise. Both of these occur as a consequence of aging. In the case of anabolic resistance to muscle protein synthesis, we are looking at what is called a "blunted" (altered, or reduced) response of muscle tissue to utilize dietary protein (and specifically, their amino acids) for MPS. What we need to keep in mind is that dietary protein is the source of new muscle tissue for our bodies when we exercise. When we are young, the body very efficiently and effectively digests, absorbs, and utilizes dietary protein and converts it into the protein-rich muscle tissue when we exercise. But with aging, this process does not occur like it used to, especially if the protein amount consumed is inadequate.

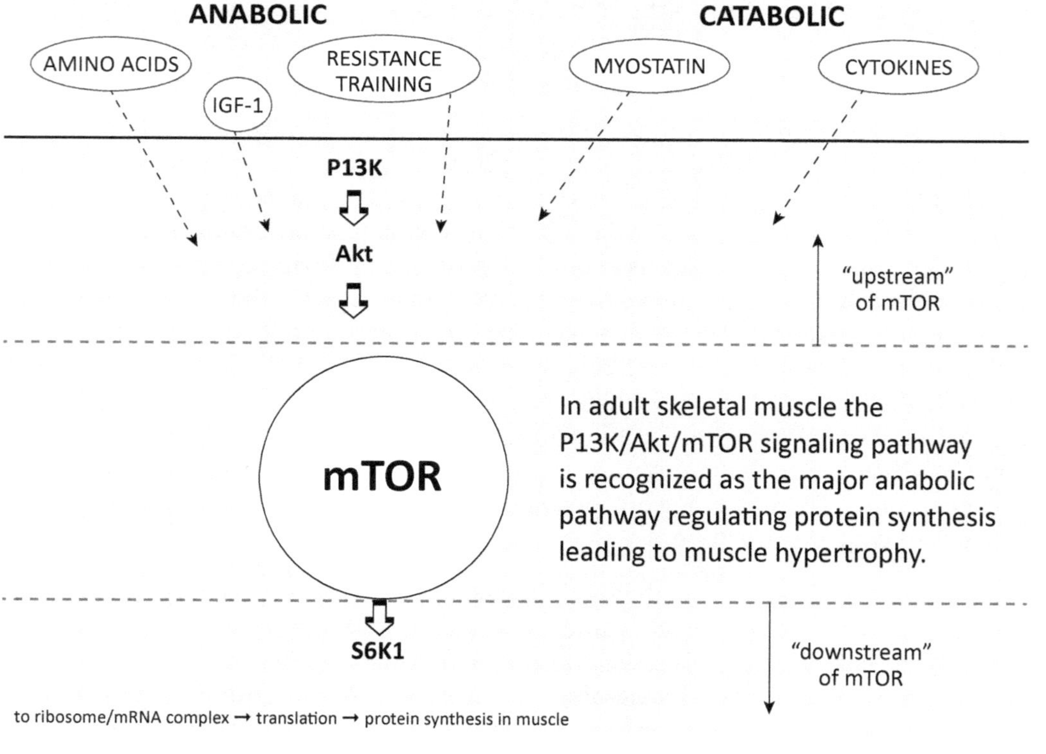

▲ **Figure 2.13** A highly simplified overview of anabolic and catabolic cell signaling processes that impact mTOR and muscle protein synthesis and hypertrophy.

At this point, I wish to briefly discuss a key molecule that is involved in protein synthesis, called **mTOR**, and which is relevant to the topic of anabolic resistance. This molecule is actually an enzyme – a kinase enzyme (transfers phosphate groups) – that controls cellular processes - including MPS. The figure 2.13 attempts to show how mTOR is sort of a master regulator of all things anabolic. It is specifically active in skeletal muscle as part of the cell signaling process that responds to nutrients (protein intake) and helps the body build muscle. As we age, unfortunately this system begins to falter. I hope I have not lost the reader due to all of this technical information, but the bottom line here is that as we age, we are less able to build muscle because we are less responsive to (a) dietary protein, and (b) anabolic exercise.

Regarding anabolic resistance to exercise, what this means is that exercise itself is a potent stimulator of muscle growth, especially when it is anabolic exercise, as we have just discussed. However, as we age, our muscle tissue is not as responsive to anabolic exercise compared to when we were younger. This means that we can expect to experience less muscle growth for the same amount of exercise. This is due to various factors common in aging men – declines in anabolic hormones (testosterone, growth hormone, and others) included. In addition, it has been discovered that the typically seen decline in physical activity, and increased sedentary behavior leading to inactivity in older age, sets the aging man up for this type of anabolic resistance.

To sum up: older skeletal muscle is less efficient at being stimulated to engage in and grow new muscle tissue by the previously responsive stimuli of (a) protein intake and (b) exercise. However, the good news is that it is possible to overcome decreased response to dietary protein by substantially increasing the protein intake, and to also overcome anabolic resistance to exercise with age-appropriate strength training. Read on and you will learn more specifics regarding addressing these challenges.

Body Changes

The human population on our planet is aging. Did you know that the number of Americans aged 65 and older is projected to increase from the current (roughly) 46 million to almost 100 million by 2060? This would mean that the number of older adults will actually exceed the number of children under age five for the first time. Longevity will increase, and it is expected that one out of every four 65-year-olds alive today will live past age 90. Nowadays you may hear people say things like, "Sixty is the new forty." This alludes to the fact that previously-held beliefs about aging are no longer valid given recent scientific advances, and changes in lifestyle (adequate exercise, good nutrition, etc.) that are possible. In plain English, a sixty-year-old man today has a better chance to look a good deal younger than his counterpart back in the 1950s.

The human body is made up of fat, lean tissue (muscles and organs), bones, and water. After age 30, people tend to lose lean tissue. Your muscles, liver, kidney, and other organs may lose some of their cells. This process of muscle loss is called muscle atrophy. Bones may lose some of their minerals and become less dense (a condition called osteopenia in the early stages and osteoporosis in the later stages). Tissue loss reduces the amount of water in your body.

The amounts and relative proportions of body fat and the lean tissue (also called fat-free mass, FFM) change as we get older, and predispose us to debility and disease. Such changes are due to of a variety of factors, including physical inactivity, hormonal status, nutrition, and illnesses. Men often gain weight until into their fifties, and may begin to lose weight in their seventies and beyond, which may be related to testosterone level changes. This later weight loss occurs because fat replaces lean muscle tissue, and fat weighs less than muscle. It is important for researchers to keep investigating these changes in order for interventions to be developed that address functional and health limitations associated with old age.

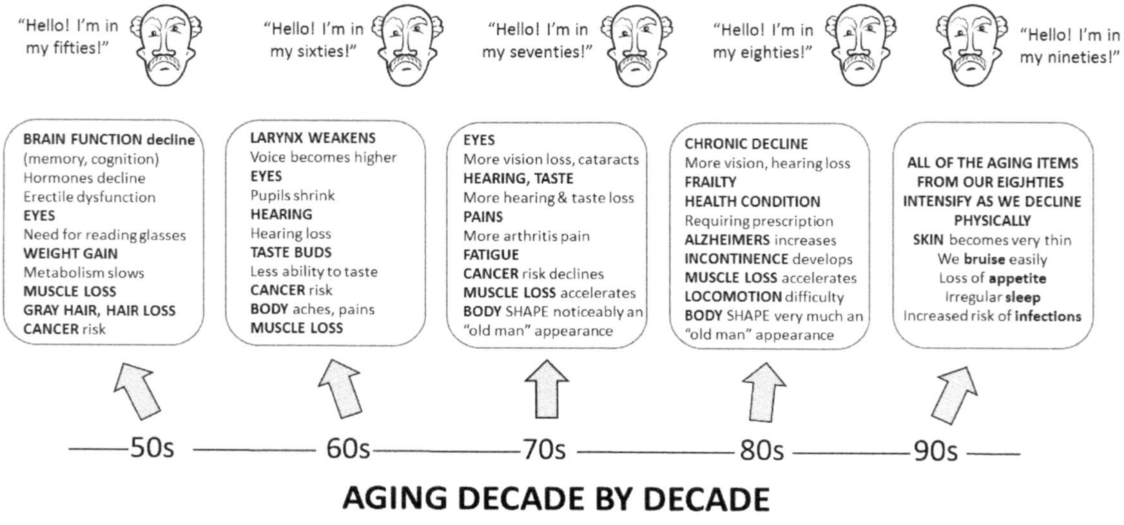

▲ **Figure 2.14** As we age, we become less youthful in appearance, as what we see when we look in the mirror is our body physically aging. Individual people age at different rates, based upon a variety of factors including genetics, lifestyle, and the factors mentioned previously. For this reason this figure is not universally applicable; however it gives an idea of changes that may be noted with the passage of time.

Frailty and Decreased Strength

I couldn't start writing this section without first recalling one of my favorite movies, 2008's "Gran Torino", starring Clint Eastwood, who I believe was 78 years old at the time of the film's release. What struck me, well there were several things – but at first, having grown up with not only the Dirty Harry films but also the classic Spaghetti Westerns and TV's "Rawhide" – well, I hardly recognized the aging Clint in his first scene in "Torino". I mean, I was expecting Dirty Harry, I guess. But we all get old. And the more I saw of him in the film, I realized he did look pretty good for his age.

But I also thought that a couple of times, his movements betrayed his years. What I mean is, he looked a bit unsteady on his feet. Don't get me wrong – at that age, you are entitled to be somewhat wobbly getting up from a chair or walking up and down stairs. And, it's certainly possible that since he was playing an old man character, he may have been "acting" that way to appear appropriately old. At any rate, it fits right in with the topic here of having less strength and being more frail as we age.

One other example from that movie… there is a scene when Clint's character Walt confronts a gang who is messing with the Asian girl. What he does is a parody I'm sure of his own devising – spitting on the ground just before he's about to take action. The original movie depicting this behavior was the 1976 "The Outlaw Josey Wales". Well that film was made thirty-two years earlier, when Mr. Eastwood was 46 years old. The difference is, in the more recent film, when Clint's character spits, the spittle has really no velocity or range – it just kind of dribbles out of his mouth to the ground. When Josey Wales spits, it's a projectile, with deadly aim! So, just another interesting reference from a film that shows an example of the decline in our older years, in this case, losing the ability to spit forcefully and with distance – as realized by one of the world's top actors.

▲ **Figure 2.15** Clint Eastwood's acting career has spanned decades.

One of the major behavioral tendencies that can be prevented in older individuals is inactivity. As old men, we are likely to have ended our careers and stopped working, at least full-time. Perhaps we were a farmer, or a policeman, or a businessman. We may have had a desk job, or were out running around all the time – but suddenly, all of that stops. This should not mean, however, that we become generally less active. Researchers refer to that as "self-selected inactivity" – when we just sit around. The problem with it is that it perpetuates anabolic resistance, with the accompanying loss of muscle mass, muscle strength, and function. This is the definition of sarcopenia, and when advanced, we become frail and at risk for falling because we lack the strength to hold ourselves upright. At this stage, we are no longer able to live totally independent lives – we become functionally dependent. Lifestyle strategies and interventions that will be discussed in upcoming chapters can help – and remaining active is one of the primary considerations.

The amounts of fat and fat-free mass (**FFM**) in adults change with increasing age as a function of a variety of factors, from physical activity to hormonal status to nutrition and disease. The situation regarding the fat and FFM compartments is associated with and serves as established risk factors for a variety of chronic diseases occurring between middle to old age. Body composition can also be used in assessments of functional status, disability, and mortality. Understanding the scope of the age-related changes in body composition and the factors associated with them in healthy adults will help to improve our knowledge and understanding of these processes and assist in the prevention of functional limitation and in the management of health status into old age.

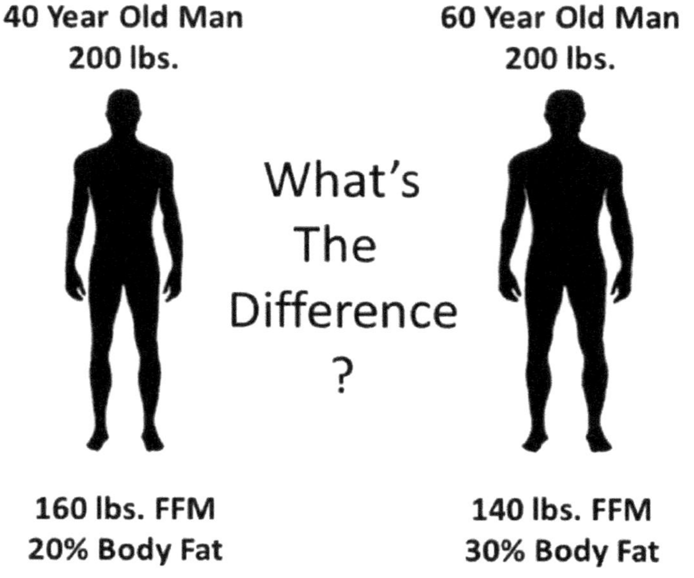

▲ **Figure 2.16** Aging and body composition in the absence of strength training. The 60 year old man (right) has a higher percentage of body fat although maintaining the same weight as when 40 years old (left). Rather than carrying around 40 lbs. of body fat, it is now at 60 lbs. This extra fat sets the stage for insulin resistance, leading to diabetes and cardiovascular disease. The extra fat also causes a decrease in testosterone and increase in estrogen. Without resistance training, his muscle (lean body mass) portion of FFM significantly decreases. A consequence of this is a lower RMR (resting metabolic rate), making him more prone to adding fat pounds. With less functional muscle mass (sarcopenia), the man is at increased risk for future functional disability. (FFM = fat free mass, which includes not only muscle, but the organs, bones, and tissues that are not fat – and of course, body water).

Sleep Inadequacy

If we were to consider three factors that most impact our ability to successfully train and develop new strength and muscle, those three would be our diet, our training program, and sleep. Did you realize that the most important of the three is… sleep? Yes, it is, because without sufficient daily sleep, it won't matter how good our diet is or our training – we will fail to execute mentally as well as physically. Try being at the top of your game without 24 or more hours of sleep! Sleep is therefore considered the key lifestyle factor contributing to our health and well-being. Inadequate sleep is what we are addressing here specifically, because that will hinder our ability to build strength and muscle, and age has something to do with it.

Science has demonstrated that the hormones which are considered anabolic (most notably, testosterone and growth hormone) actually perform much of their magic during periods of overnight sleep. There is also such a thing as a sleep cycle, and when we sleep, we need to get full, complete sleep cycles in order to feel refreshed, and for the anabolic effects to occur. So, we need to understand a little bit about this amazing z-z-z phenomenon that we take so much for granted.

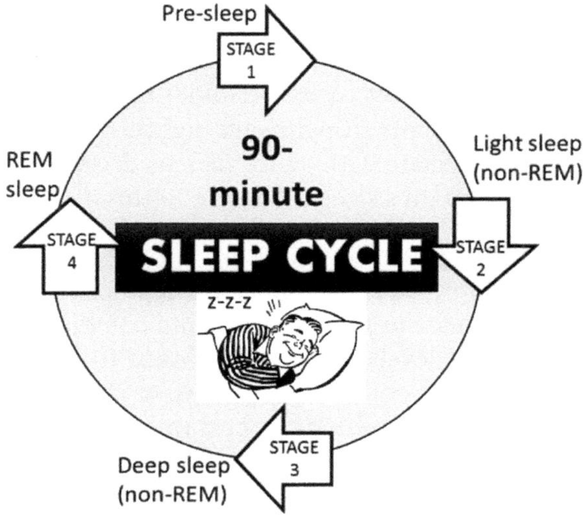

▲ **Figure 2.17** A healthy sleep cycle lasts for about 90 minutes. This means that for sleep totaling 7.5 hours we cycle five times, which is considered healthy. Four or fewer complete cycles are inadequate by comparison. The sleep cycle is actually a bit more complex than as presented, because throughout the night, the time spent in each stage will shift. For example, during the first few 90 minute sleep cycles, we spend most of our time in deep non-REM sleep (stage 3), but during the later sleep cycles as we approach morning, we spend comparatively more time in REM sleep. Age also is a factor; the older we are, the less deep and REM sleep we experience.

There are four stages in a complete sleep cycle, and it is to our advantage to know something about each. The first stage is sort of pre-sleep, when we are just starting to doze off. We then move quickly into the second stage, in which the heart rate and brain activity slow down. It is somewhat difficult to be awakened from Stage 2 sleep. In the third stage of sleep, important events occur. This stage is referred to as "deep sleep", and it is here that the anabolic hormone secretions take place. Also, the brain produces very slow waves known as delta waves in this stage. It is also known that if a sleeper is awakened during Stage 3, they are often disoriented. Sometimes this stage is subdivided into deep and "very deep", adding an extra stage to make the total number of stages five (however we will keep it simple with four). The fourth and final stage of sleep is REM sleep (rapid eye movement). You may have noticed your pet dog or cat in REM sleep – the eyes twitching, heavy breathing, whimpering and so forth. Well, during human Phase 4 sleep, we show the same response as our furry friends. Our dreams also belong to this stage (yes, dogs and cats dream too) and we do become temporarily paralyzed at this time. You are likely familiar with the frustrating situation during sleep when you feel the need to shout but cannot – this is because of the temporary paralysis (fortunately we do come out of it).

With Stage 4, we have completed one full cycle, and then return to Stage 1 for another cycle, and then another and so forth until we finally awaken. So you can see how it is possible to have multiple dreams throughout the night of sleep, in accordance with each complete cycle. There are also multiple anabolic sessions occurring during which time growth hormone and testosterone are actively functioning throughout our sleeping bodies, helping with cellular restoration and rebuilding in preparation for the new day. The take-home message here is that if we skimp on sleep, we cannot adequately build muscle.

As hinted at earlier, as we age, we potentially face a couple of sleep-related challenges. One is that we spend less time during each sleep cycle in deep sleep. Young men are in Stage 3 about 20% of the night, but this reduces to 10% in older men (aged 40s and 50s), and gets into the single digits – as low as 2% - by roughly age eighty. It has been suggested that some of the characteristics that we associate with aging such as decreased mental acuity and lack of feeling refreshed after a night's sleep – are due to this decline in the amount of Stage 3 sleep.

The other challenge is that we don't get enough total sleep hours at night, in part because we don't assign as much importance to this aspect of life compared to others during our waking hours. Lower testosterone levels may also be linked to fewer sleep hours in men. As already implied, sleep deprivation is counter-productive to strength training and building muscle. Therefore it is imperative that we try our best to get sufficient quality and quantity of sleep each night. Some workaholic men out there may say "I don't have time for sleep!" – but what a mistake that would be. Another phrase that is humorous, which I have heard spoken, is "I can catch up on my sleep when I'm dead" - ! The truth is, the sooner we realize how significant sleep is to our strength, health and overall well-being, the sooner we will realize that the better we sleep, the better we will train, recover, grow, and function. We even burn fat more efficiently with a good night's sleep. Bottom line: we must become as devoted to sleep as we are to training – the two complement and support each other, like bread and butter.

SLEEP BENEFITS	
Allows for anabolic hormone secretion	Helps build muscle
Boosts mood and energy	Helps control hunger
Burns calories	Improves heart health
Fixes brain circuitry	Maintains immune system functioning

▲ **Table 2.4** Notice all the listed benefits of sleep. That's a lot of good stuff for virtually every aspect of a man's health.

Research has shown that sleep is vital for recovery following a workout, and aids with muscle repair. Said the other way around, inadequate sleep hinders the body's ability to recover after strength training and also interferes with building maximum muscle strength. During stage 3 sleep, blood flow to the muscles increases, allowing for muscle tissue growth and repair. During REM sleep (stage 4, prior to waking), our muscles relax. This helps relieve body tension and may reduce chronic pain, and hasten the muscle-healing process if we injure ourselves. Did you know that our muscle coordination improves when we have good sleep habits? Studies have shown increases in reaction time and speed which allows for overall improved performance – athletic, or simply functional. Do you have trouble sleeping? Strength training may help achieve a better night's rest by reducing stress and frankly, making us tired. This translates to falling asleep faster, having a more restful sleep, and waking up less frequently throughout the night.

Low Testosterone

I've already mentioned a link between testosterone production and sleep. But we have the problem of experiencing decreased testosterone levels not just when our sleep is off. It also happens routinely to men as they age. Of relevance to us as we begin our strength training, research has shown that the older body slows down production of the two key anabolic hormones - testosterone and growth hormone - that are beneficial to working out. Performance in the gym, as well as on the field or the court for you weekend warriors, takes a nose dive for this reason. Building muscle and staying lean is much easier when we are young due to having robust hormone levels and response, and testosterone is well-known as the primary male sex hormone. Although key, sex drive isn't the only function that testosterone performs; it is actually quite an all-star hormone, responsible for such diverse effects as a sharp mind, improving our mood, developing stronger bones, building muscle, and keeping body fat in check. Although declining "T levels" is common to aging men, we will learn that certain strength training exercises, in addition to other lifestyle choices, can help preserve and potentially boost this hormone – even in our later years. In fact, any time you see a fellow who is sixty or older sporting a muscular physique (assuming they are not taking anabolic steroids), you can assume that their lifestyle choices have enabled them to maintain their T levels naturally and safely.

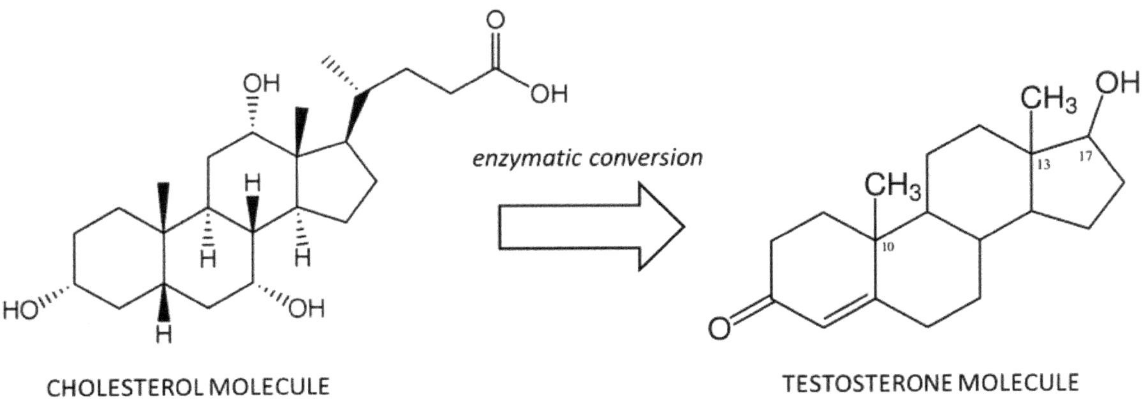

▲ **Figure 2.18** The testosterone molecule is synthesized from its precursor, cholesterol. Notice the similarity between testosterone structure compared to cholesterol structure. In the Leydig cells of the testes, cholesterol is converted into testosterone by an enzymatic reaction triggered by lutenizing hormone, which is responsive to growth hormone secretion. So, cholesterol is the building block of testosterone, and the correct interplay of hormone activity is essential. The overall process is rather complicated, since both the hypothalamus and the pituitary gland as well as the testes are involved.

Some readers may be wondering about getting their T levels checked. As a result, I want to mention the fact that there are several different types of testosterone that exist in the body. These are **free testosterone, SHBG-bound testosterone** (Sex hormone-binding globulin or sex steroid-binding globulin, a glycoprotein that binds to the two sex hormones androgen and estrogen), and **albumin-bound testosterone**. The bioactive form is the free testosterone, which means that if the level of free T is low, you should discuss strategies with your physician, who may recommend some form of T therapy. Also, there is **total testosterone**, which often is assayed – but remember, it is the free form that matters most.

Of interest, the occurrence of low testosterone in men as they age is typically accompanied by a rise in the opposing hormone, estrogen. For women, the opposite situation may be true. When I was a youngster, I used to think that as men aged, they began to look like "old aunties", while the women became more masculine. I also heard that older couples began to more closely resemble each other physically. Although speculation, the decline in each gender's primary hormone would make this appear to be possible. Some consumers have heard about "estrogen blockers" and are actively taking these as supplements. Is this prudent? Estrogen blockers (aromatase inhibitors) have been prescribed in women who have breast cancer to lower estrogen levels, while some men may want to use them to retain testosterone, but these drugs are not without side effects.

Low Vitamin D

When I visited my doctor for a check-up at age 63, my blood was analyzed, and it revealed a Vitamin D deficiency. I was surprised, and began to take Vitamin D supplements as recommended by the doc – and spent more outdoor time in the sun. It turns out that this kind of deficiency in older men is not at all uncommon, but it is of concern. A deficiency of this vitamin is implicated in osteopenia, osteoporosis and the bone fractures and falls to which elderly persons are prone to.

Technically referred to as Vitamin D3, **cholecalciferol** or simply D3. Vitamin D is somewhat difficult to obtain by dietary means. This is one reason milk is fortified with this vitamin – although it is a fact that milk consumption in the USA has been on the decline for years. I saw this firsthand when I visited elementary, middle and high schools as a nutrition researcher. I was amazed at how many unopened cartons of milk were returned by the boys and girls at the end of the lunch period. However, D3 can be synthesized in the skin by the action of the sun's ultra-violet rays – so being outdoors helps when dietary intake is poor. But if we tend to remain indoors, as is typical in older adults, we are likely to be deficient, so supplementation becomes a worthwhile and needed strategy.

BIG CHALLENGE: SARCOPENIA

I have been addressing what were identified by me as "consequences of aging", and I would be totally remiss if I failed to highlight one of our biggest challenges as old men: it is called sarcopenia. Are you aware of it and know what it is? If not, brace yourself. It is important enough that I have moved it out of the section on aging consequences so that it becomes the star of its own section of the chapter, and doesn't get lost in the shuffle.

Here's the deal. No matter who you are, and how athletic you may have been, we all lose muscle mass to a certain extent as we age, and this means "net". In other words, we build less muscle daily than we lose, making us less muscular and less functional. And this deficit occurs continuously: daily, weekly, monthly. Typically it starts happening to us when we are in our 30s, at about 1% per year, but it accelerates such that when we reach our 70s and 80s, we may have lost up to 40-50% of our muscle mass. It is due to normal daily protein turnover, wherein muscle protein loss (breakdown) exceeds net muscle protein formation (synthesis).

This condition, sarcopenia, is critically important to be aware of due to the fact that we can lose our functional independence as a result of the loss of strength as we lose muscle. A consequence of sarcopenia is that we experience weakness, frailty and a lower **QoL** (quality of life). To be clear, sarcopenia is characterized by decreases in both muscle mass and muscle function. As stated, this situation is caused by an imbalance between the rates of muscle protein synthesis (anabolism) and muscle protein breakdown (catabolism) such that a net negative balance exists:

muscle protein synthesis (MPS) < muscle protein breakdown (MPB)

which can also be represented this way:

muscle protein breakdown (MPB) > muscle protein synthesis (MPS)

Thus, muscle mass is maintained by the parallel and opposing processes of **MPS** and **MPB**. It is thought that increases or decreases in muscle mass are caused by ongoing changes in muscle protein synthesis rates, breakdown rates, or a combination of both.

It is generally accepted that alterations in MPS are regulated by what are termed anabolic stimuli, which includes food intake (especially protein) and physical activity, especially resistance training. Dietary protein not only increases MPS rates but also inhibits MPB, producing net muscle protein gain. Resistance training is a bit different. Lifting weights provides the needed stimulus to cause muscle tissue to grow, but this only happens if protein is available after the initial damage and MPB caused by the workout. Notice – working out contributes to more MPB than if we don't exercise! That is why our nutrition (as discussed in Chapter 5) is so important. And the best way to overcome sarcopenia is to engage in strength training and consume an adequate amount of the right kind of protein – and this amount is likely more than what you are currently ingesting.

Now for an unpleasant reality. It turns out that as we age, we develop a blunted MPS response to anabolic stimuli. This means that we exhibit less muscle growth even though we present ourselves with resistance stimulus by working out with weights. A similar blunting occurs with respect to the dietary protein stimulus. Such effects are referred to by the experts as anabolic resistance of aging, which happens to folks in our age group of 50-90-plus years old. The older we are, the worse it gets. So you see, we have two strikes against us! The good news is, it is believed that regular resistance training and increasing the dietary protein can restore the sensitivity of older muscle to MPS, thus overcoming anabolic resistance. H-E-L-L-O! We covered this earlier – just checking!

By the way, the decreased MPS response to resistance exercise in the aged adult population compared to the young is thought to be due to decreased anabolic hormones, and less efficient and less effective anabolic cell signaling (recall mTOR). Fortunately, there are ways to train and lifestyle adjustments that we can make that help boost, for instance, our declining testosterone levels. Importantly, there is also the strategy to indirectly boost MPS by doing whatever we can to decrease the amount of muscle catabolism (MPB) through diet and lifestyle. This works because if we experience less MPB, then MPS will tend to dominate MPB (MPS > MPB), and we will gain muscle mass over time rather than lose it.

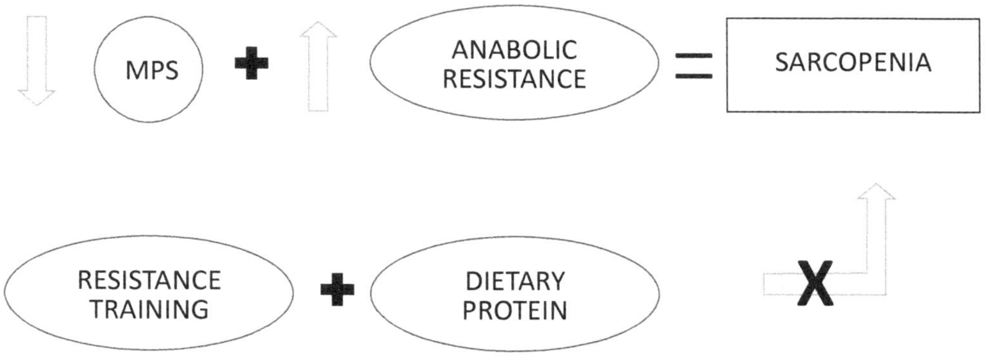

▲ **Figure 2.19** The preservation of lean muscle mass and functional capacity are key components of healthy aging and the maintenance of independence. Top, left to right: MPS (muscle protein synthesis) declines with aging, and when combined with increased anabolic resistance, results in sarcopenia. Research demonstrates that MPS does not decrease and sarcopenia does not result when intake of sufficient protein and resistance exercise are simultaneously practiced.

Having discussed serious issues of aging including anabolic resistance and sarcopenia, we'll now conclude this chapter on a somewhat lighter note, examining what is called "**Grumpy Old Man Syndrome**".

GRUMPY OLD MAN SYNDROME

Another consequence of aging is that of becoming a GOM: a grumpy old man. Although there are many stereotypes on television (Archie Bunker, Perry White and Oscar Madison), movies (the 1993 film called "Grumpy Old Men" starring Walter Matthau and Jack Lemon), and daily (family and work) life - I turn to actor Clint Eastwood and his portrayal of Walt Kowalski in the film "Gran Torino". Now, you knew that this character was indeed a grumpy old man when he growled "Get off my lawn" in that movie. But what is it that tends to make old men grumpy? Let's look a little deeper into GOMs, or as it is called in the medical arena, IMS (irritable male syndrome).

According to author Michael Gurian, an expert in male psychology, hormonal changes are partly to blame. The decline in testosterone makes us less than we once were, and that makes us MAD. So, we become irritable and grouchy. But there is more to it – old men who are overweight and don't exercise tend to be more irritable because they lose any athletic identity they may have had when younger. And then, there is the reality of sarcopenia to deal with…

Gurian, who also acknowledges the damage caused by losing our work identity upon retirement leading to grumpiness, does offer some advice for us. First, we must let go of our lost youth, as it is really and truly gone and is not coming back. Instead, we all need to create new, more age-appropriate identities for ourselves, and I suggest that we do so by becoming Old Man Muscle workout junkies. Similarly, Gurian advises us to shift our energies into something new. I can see that for many men, this could mean not only their physical but also their spiritual development. And this is this is why I have written the sixth chapter this handbook on spiritual strength training – so please look for and take note of it.

Gurian feels that every man needs to make these kinds of shifts somewhere between the ages of fifty and sixty-five. By then, we will no longer be anchored in "what I was", but instead we reinvent ourselves into the new "what I am now becoming". And if you are already over sixty-five, sure, it is okay to get started at whatever your age. What we are discussing here is a renewal process utilizing available energy and vitality that will work wonders for our psyche – which is the opposite of the decline and decay associated with grumpiness. Perhaps with sufficient effort, the grumpy old man can revert to become the happy old man after all! I'm willing to try, how about you?

▲ **Figure 2.20** Jack Klugman's television character Oscar Madison was quite a grump and a grouch.

"Grumpy Old Man Syndrome"	
• We feel less attractive to the opposite sex • The opposite sex seems less attracted to us • We may become divorced or lose our spouse • We have less muscle (and fewer teeth) than ever • We can't move, see, hear or remember like we used to • There are changes to family (kids grow up and move out) • Body and joint aches and pains that accompany older age crop up • We may have trouble with digestion, with sleep, with incontinence	• There are changes to the world around us (that we don't agree with) • There are changes to work life (we become unemployed and/or retire) • Our physical appearance declines (we get shorter, flabby, wrinkled, lose hair) • We lose any sports/athletic ability that we may had in our younger days • As GOMs we may begin to enjoy being grumpy to people – our new identity!

▲ **Table 2.5** While "grumpy old man" is a common expression, there is an actual medical term for this condition (or as it is called, syndrome), which is "Irritable Male Syndrome".

RECAP AND WHAT'S AHEAD

With this section on grumpy old men, we close out our second chapter. We now have a better idea of the challenges of aging, as well as the possibility to address them in a positive and practical manner. We can answer the question: why should we get involved with strength training? Our answer: in order to address the multitude of health declines associated with aging, not the least of which is sarcopenia, in order that we have a chance to decrease lifestyle disease risk (diabetes, obesity, cardiovascular disease, cancer) and maintain good health during our later years. Secondly – to become stronger, more fit, and improve our physique in order that we have a chance to maintain a higher functional quality of life during our later years. So, it is about developing muscles, yes, but it is so much more: it is to improve our overall health and quality of life during our older years.

It should now be clear that the benefits of physical activity/strength training to us as old men go beyond gaining strength and building muscle. Strength training can improve our cardiovascular and bone health, help lower our blood pressure, improve our insulin sensitivity to achieve normal blood glucose levels (as A1c), assist with weight loss, provide a mental and energy boost, improve our body composition, boost testosterone levels, and decrease the risk of becoming frail and falling. What's more, the benefits of strength training have been well-documented in the research literature even for persons into their tenth decade of life. Furthermore, I suggest that strength training, coupled to spiritual training and the dietary (food choice) strategies outlined later in this book, can make it possible for each one of us to become a new person that gets through the aging process in a more positive manner. What great news – especially since not many people attempt to reinvent themselves at an older age – they instead accept the decline and just go with it. But I suggest that becoming a revitalized older person should be our goal - one having more vitality and "life force", possessing a new kind of happiness that makes life still very much worth living.

Coming up next in Chapter 3 is an introduction to the **exercise science** of strength training, which represents good background information for us as we prepare to begin our Old Man Muscle strength training program. I recommend that everyone read it before putting on your gym attire and diving into the chapter that details the Old Man Muscle workout.

RESOURCES

Armanios M, Blackburn EH. 2012. The Telomere Syndrome. Nature Reviews. Genetics 13(10): 693–704.

Baechle T and Earle R. 2008. Essentials of Strength and Conditioning, 3rd edition. Champaign, Ill. Human Kinetics.

Blackburn EH, Epel E. 2017. The Telomere Effect: A Revolutionary Approach to Living Younger, Healthier, Longer. Grand Central Publishing.

Born J, Muth S, Fehm HL, 1988. The significance of sleep onset and slow wave sleep for nocturnal release of growth hormone (GH) and cortisol. Psychoneuroendocrinology 13(3):233-43.

Breen L, Phillips SM. 2011. Skeletal muscle protein metabolism in the elderly: Interventions to counteract the 'anabolic resistance' of ageing. Nutr Metab (Lond) 8:68 Accessed at: https://nutritionandmetabolism.biomedcentral.com/articles/10.1186/1743-7075-8-68?optIn=true#citeas

Bryant CX, Jo S, Green DJ. 2014. Personal trainer manual. American Council on Exercise, 5th ed.

Bryant CX, Jo S, Green DJ. 2014. ACE essentials of exercise science for fitness professionals. American Council on Exercise.

Clements SJ and Carding SR. 2016. Can nutritional intervention counteract immunosenescence in the elderly? In: Molecular Basis of Nutrition and Aging, A Volume in the Molecular Nutrition Series. 375-391.

Gokhale R, Chandrashekara S, Vasanthakumar KC. 2007. Cytokine response to strenuous exercise in athletes and non-athletes-an adaptive response. Cytokine. 40:123–127.

Gomez-Cabrera MC, Domenech E, Viña J. 2008. Moderate exercise is an antioxidant: Upregulation of antioxidant genes by training. Free Radic Biol Med. 44:126–131.

Green DJ, Bryant CX. 2017. Personal trainer manual. American Council on Exercise, 5th ed.

Gurian M. 2013. The wonder of aging: a new approach to embracing life after fifty. Atria Books/Simon and Schuster.

Hatfield F C. 2019. ISSA Fitness: The complete guide. The International Sports Science Association, 9th ed.

Janssen I, Ross R: 2005. Linking age-related changes in skeletal muscle mass and composition with metabolism and disease. J Nutr Health Aging 9: 408-419.

Jones CJ, Rose DJ. 2005. Physical activity instruction of older adults. Human Kinetics.

Jones MJ, Goodman SJ, Kobor MS. 2015. DNA methylation and healthy human aging. Aging Cell 14(6):924-932.

Jung M, Pfeifer GP. 2015. Aging and DNA methylation. BMC Biology 13:7.

Karacan L et al. 1974. Growth hormone levels during morning and afternoon naps. Behav Neuropsychiatry 6(1-12):67-70.

Kreider RB. 2019. Essentials of exercise & sport nutrition: science to practice. Lulu Publishing Services.

Marcell TJ. 2003. Sarcopenia: Causes, Consequences, and Preventions. The Journals of Gerontology Series A. 58(10):M911–M916.

Minson M. 2012. Prepare to defend yourself… how to age gracefully & escape with your dignity. Texas A&M University Press.

Narici MV, Maffulli N. 2010. Sarcopenia: characteristics, mechanisms and functional significance. Br Med Bull. 2010, 95: 139-159

O'Brien CS. 1998. Exercise, aging and health: overcoming barriers to an active old age. Taylor & Francis. Philadelphia, PA.

Ong CKS, Lirk P, Tan CH, Seymour RA. 2007. An Evidence-Based Update on Nonsteroidal Anti-Inflammatory Drugs. Clin Med Res. 5(1): 19–34.

Pae M, Meydani SN, Wu D. 2012. The Role of Nutrition in Enhancing Immunity in Aging. Aging and Disease. 3(1):91-129.

Pilz S, et al. 2011. Effect of vitamin D supplementation on testosterone levels in men. Hormone and Metabolic Research 43(3):223-5.

Reynolds G. 2012. The first 20 minutes. Icon Book Ltd. London.

Shah K, Villareal DT. Obesity. In: Fillit HM, Rockword K, Young J, eds. 2017. Brocklehurst's Textbook of Geriatric Medicine. 8th ed. Philadelphia, PA: Elsevier; Chapter 80.

Simioni C, Zauli G, Martelli AM, et al. 2018. Oxidative stress: role of physical exercise and antioxidant nutraceuticals in adulthood and aging. Oncotarget. 9(24):17181-17198.

Uchida MC, Nosaka K, Ugrinowitsch C, Yamashita A, Martins E, Jr, Moriscot AS, Aoki MS. 2009. Effect of bench press exercise intensity on muscle soreness and inflammatory mediators. J Sports Sci. 27:499–507.

Van Cauter E, Plat L. Physiology of growth hormone secretion during sleep. J Pediatr. 128 (5 Pt 2):S32-7.

Walston JD. 2016. Common clinical sequelae of aging. In: Goldman L, Schafer AI, eds. Goldman's Cecil Medicine. 25th ed. Philadelphia, PA: Elsevier Saunders. Chapter 25.

Westcott WL. 2015. Build muscle, improve health: Benefits associated with resistance exercise. ACSM's Health & Fitness Journal 19(4):22–27.

Zatsiorsky V, Kraemer W. 2006. Science and Practice of Strength Training, 2nd edition. Champaign, Ill.: Human Kinetics.

BEFORE YOU START TRAINING

Introduction

We have just finished a chapter discussing the toll that aging takes on our bodies, and have already hinted at the possibility that strength training can be an effective anti-aging weapon. In Chapter 3 we will examine some of the latest information coming from the scientific community regarding what happens when we engage in strength training. Now, there are many potential ways to strength train, just as there are many exercises to choose from. I believe that, based upon what has been written by others as well as my own recent experience as an older man, for most individuals reading this book, working out with dumbbells and other free weights (barbells, kettlebells) during the first year is the gold standard.

The older man should never start working out without the assurance that he is physically able to begin. An appointment with a physician to check out our health is in order. This includes consideration of our age, the presence of any cardiovascular, pulmonary and cardiorespiratory risk factors, metabolic risk factors, family medical history (parents, other blood relatives), and so forth in order to determine if we can be cleared to begin a strength training exercise program.

Some readers of a more advanced age, or previously untrained, perhaps very out of shape, or having an injury or a pre-existing health condition, really need to have a professional assessment performed by a certified fitness professional prior to any recommendation regarding what, when, and if is the best way for that particular person to begin training. In such instances, machines that use pulleys and cables may be preferable to free weights at the start, as they may provide less stress on the joints. However, in most cases, the correct dumbbell weight when properly selected and coupled to specific exercises for even untrained elderly individuals can provide superior training and health results. Supervision by a trainer in such cases is always a good idea. Yes, it costs some money to hire a trainer – but since we are talking about an activity that can serve as powerful medicine to make us healthier, potentially live longer, and be stronger, it is money well-spent.

Readers should recall from Chapter 2 the distinction between training for muscular endurance as compared to training for muscular strength. While muscular endurance can be increased by performing more repetitions with a particular resistance, gains in strength come as a result of subjecting muscles to progressively heavier training loads. This is called **muscle overload**. Gains in muscle size are linked to increases in overall training volume over time.

I would just add at this point that while it is often stated from the physiological perspective to train in such a manner so as to maximize results via constantly changing the workout by increasing the amount of weight and number of repetitions, as well as the exercises themselves, this is not always necessary. In other words, when not training for a sport as a competitive athlete or bodybuilder, with all of their inherent demands, it is fine to train recreationally using a fairly consistent and less stressful workout routine and schedule in order to achieve and maintain a certain level of strength and muscularity. Remember, just the fact of performing any exercise at all on a consistent basis is very positive in terms of the potential health benefits compared to no exercise. As older men who are training recreationally rather than in a competitive, all-out manner, we can do quite well with a plan that has some elements of change and progression built into the workout over time but without changing things up too frequently, aggressively and forcefully.

It is important when devising any strength training program to keep in mind the parameters that will result in success. These can be categorized as follows: (1) work the muscles against loads sufficient to stimulate growth but do so in an age-appropriate manner; (2) do not over train by allowing for sufficient rest between exercise days (typically, train only two or three times per week depending on factors including age, training background, and overall physical condition); (3) follow proper form for each exercise, practicing good biomechanics, which means allowing the muscles to move in the manner they were designed to move; (4) support training with a proper diet that complements muscle repair and growth; and (5) follow sleep patterns that allow full restoration to take place each and every day. The latter requires in general a minimum of 6-7 hours of sleep, and may be supplemented with daily "power naps" or, as I call them, "old man naps". Training the Old Man Muscle way seeks to employ all of these components.

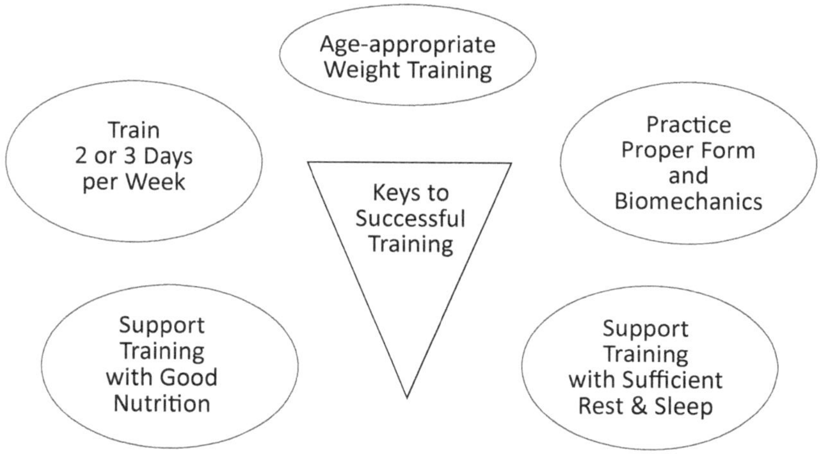

▲ **Figure 3.1** Five principles to follow when strength training: (1) work the muscles against loads sufficient to stimulate growth but do so in an age-appropriate manner; (2) do not over train - allow for sufficient rest between exercise days (typically, train only two or three times per week depending on factors including age, training background, and overall physical condition); (3) follow proper form for each exercise, practicing good biomechanics, which means allowing the muscles to move in the manner they were designed to move; (4) support training with a proper diet that complements muscle repair and growth; and (5) follow sleep patterns that allow full restoration to take place each and every day.

We will get to the topic of exercise science in a jiff, but there's something I must get off my chest first. And it is this. Wouldn't it be great if after a handful of strength training workouts we could see muscle growth in our arms, legs, chest and so forth, and then quit training because we like the results and expect that our muscles will now look this way from now on? Unfortunately, this is not the way things work. Strength training must become a regular lifestyle event for us from now on, for as long as we can, if we are to see physical results and reap the many possible health benefits. Even for those younger men decades our junior, the effects of training on improving our strength and physique take time.

One of the realities that will become obvious to anyone who trains is that muscle does not last. What I mean is, when we train for strength, we are also building muscle tissue, and this will become visibly obvious to a certain degree. But unlike fat tissue (which seems quite happy to persist and be resistant to loss), muscle tissue "disappears" quite easily and rather quickly soon after we stop training. And it seems to be more so the case as we age. If we stop working out regularly, we shrink. If we get the flu and are out of commission for a week or more, we may feel like we have lost all of our hard-fought gains! This can be very frustrating, but when we know in advance to watch out for it, we can to an extent prevent or at least limit what I call **VMS**, or "vanishing muscle syndrome". The best approach is to train consistently, and to have enough protein in the diet for protein synthesis, which is the body's way of building muscle after a bout of strength training exercises. Of course we also need enough rest and sleep to allow the body to recover from our workouts and for the muscles to rebuild, or as the scientists say, remodel. And then, there is the science part, and what happens regarding what are known as myonuclei and satellite cells – thank goodness for them, I say. For more on that, see the section coming up on muscle hypertrophy.

To fully appreciate what exercise science is about requires knowledge of human anatomy, exercise physiology, biochemistry, molecular biology and genetics, and principles of biomechanics and kinesiology. Though well beyond the scope of this little handbook to cover all that is known in these areas, a few essentials will be presented.

DO I NEED A PERSONAL TRAINER?

This is a key question to consider. I have already mentioned the importance of considering training with a professional strength training coach. These folks are trained in client assessment through the administration of health history interviews, and in performing what is called **risk stratification** determination. Furthermore, certified personal trainers are skilled at obtaining and evaluating key client lifestyle information. This is important in creating the best targeted exercise program that will result in adherence and best results obtained by the client.

Something else that is important to consider before we start training: we need our own personal preparation. This means that we prepare ourselves mentally (thinking about and focusing on the workout exercises), and also that we prepare our joints, nerves and muscles. What? Yes, that's right. First warming up, and then some dynamic stretching – always in that order, does the job. Probably best without music playing – we need to focus our minds on the job at hand without distractions, and develop the all-important mind:muscle connection. More on mind:muscle later – by the way, it is not the same thing as mind: body exercise.

In many cases, as we age, we develop poor posture, joint problems, and muscle imbalances from decades of use, disuse, and/or abuse. These serve as barriers to obtaining good training results. Again, if within your means, you can consider hiring a personal trainer to help you evaluate your physical status in this regard. A good trainer knows how to assess what is called the body's **vertical alignment** or static posture, noting any deviations from normal. The trainer is also skilled in observing any muscle tightness or movement compensations indicative of faulty neural control and muscle imbalance at joint positions. If present, it is important to resolve such issues in order to straighten the body before attempting to strengthen it. The trainer will know certain stretches and exercises that can address any imbalances. In addition, a trainer will of course design a fitness program to help you meet your strength and fitness goals. A trainer might seem a luxury, but as was said earlier: consider what is at stake – the quality of our life during our closing decades here on earth.

A good trainer will provide social support and will motivate their clients with respect to working out and to self-regulate their lifestyle behaviors (proper diet, proper rest, etc.) The trainer will also provide real-time instruction and corrections to improper form and technique during exercising, as well as monitor client workout progress. Such oversight is invaluable since it can prevent injury and progress stagnation, and lead to positive strength and muscle gains plus the many health benefits associated with exercising.

<table>
<tr><th colspan="2">A good trainer will provide the following services:</th></tr>
<tr><td>• Assess client baseline health and fitness status, and suitability to begin an exercise program.
• Instruct client on correcting muscle imbalances and postural deviations prior to beginning a workout program.
• Refer clients to other healthcare professionals when appropriate.
• Develop and implement structured exercise programs in small-group settings that are safe, effective, and age-appropriate.
• Assist clients with establishing and achieving realistic fitness goals.</td><td>• Provide consultations for instructional and motivational purposes.
• Educate clients on health and fitness-related topics so these can be adopted and become part of a new and improved client lifestyle.
• Instruct clients on proper exercise form and technique to maximize results and minimize risk of injury.
• Monitor client status before, during, and after each workout to prevent overtraining and to identify any physiological deviations from normal.
• Is prepared for emergencies and responds appropriately if they occur.</td></tr>
</table>

▲ **Table 3.1** Some of the reasons why you might consider hiring a personal trainer.

I appreciate viewing on occasion the You Tube channel of former bodybuilder and wrestler Ric Drasin, and the guests and topics that he programs. Ric offers a wealth of insight and advice on training and lifestyle based upon years of personal experience, and engages guests who also share their expertise. All men should look him up on the internet. One guest that has been featured played the role of the big Swede character in the Clint Eastwood film, Heartbreak Ridge. I am talking about 6'6" actor and former NFL football player Pete Koch. He has been lifting since his teens, and at the time of the interview was 55 years old and in great shape. Ric quipped that he looks like Richard Gere "on juice" (steroids), although he appears to be a natural bodybuilder. I just wanted to summarize some of the comments he made because they support what we are trying to achieve here in the Old Man Muscle handbook.

Mr. Koch, who works as a personal fitness coach, suggests challenging the whole body (via deadlifts, squats, etc.) in addition to specific body parts (focusing on the arms, legs, chest, etc.) regularly. The issue is the quality of life as we age. He emphasized that working out should be done by older men to regain flexibility, strength, and locomotion – things that we lose as we get older. One example he mentioned is the ability (don't laugh) to get up off the toilet. This requires strength and flexibility in the ankles, knees, hips, quads, and even core. Realize how true - especially since there is often nothing to grab onto when raising up. If a person needs to develop in these areas but has no weights, Koch recommends performing "air squats" with your body weight. These are surprisingly challenging and effective. I would add that air squats can be done over a chair, and you can even simply sit and stand repeatedly using the chair for safety as a form of this exercise.

Another point made by Mr. Koch is that our workout routine – the specific layout of exercise reps and sets – when coupled to a sound nutrition plan, fit together like the pieces of our own personal healthy aging puzzle. Taken as a whole, they combine to advance our physical status so that we can grow more muscular and stronger, rather than the opposite, as the years go by. He also pointed out that many times his clients have pre-existing injuries, pains and other limitations – but these should not deter them from working out to the extent that they can. In fact, a good trainer can provide modifications of standard exercises and lifts to accommodate such situations so that practically any person of any age can regularly work out in such a manner as to build muscle and strength. Are you getting motivated yet?

Now, whether you hire a personal trainer or decide to train by yourself or with a buddy by applying the information contained in this handbook, it is important that we set age-appropriate goals in addition to applicable training strategies for ourselves. What this means in part is that our goals must be realistic and achievable for a typical person in our age group. Setting goals prior to starting training facilitates behavior change, and can be considered as a lifestyle component. Goals should follow what is called the **SMART goal** system. This acronym stands for: S (Specific), M (Measurable), A (Attainable), R (Relevant) and T (Time bound).

Strength Training Goals	
"I want to lose weight and gain muscle" "I want to tone up and be more flexible" "I want to improve my posture and increase my strength" "I want to complete every workout during phase one of year one" "After three months, I will be more flexible and will have more strength"	"I will be able to feel better about my physique and not be embarrassed when I take my shirt off at the beach" "I will be able to lift more weight per exercise at 6 months and again at 12 months of year one compared to the start" "I will complete as many workouts as possible during year one, and see positive changes in my weight and muscle mass"

▲ **Table 3.2** Several examples of strength training goals. Which ones reflect the SMART approach?

Again, a good trainer can assist the client with this very important activity. The trainer also knows the following: (1) setting too many goals is a bad idea and can be overwhelming;

(2) it is better to use positive words ("I will do this or that") rather than negative words ("I won't do this or that") when expressing one's personal goals relative to strength training in order to focus on achievement rather than avoidance; (3) set both short (each workout) and long-term (year one of training) goals; (4) set **process goals** (i.e. completing all the weekly workouts) and also **performance goals** (i.e. pounds lost, or amount of weight lifted); and (5) goals should periodically be reviewed and adjusted as needed in order to avoid discouragement and to promote positive effort, persistence and satisfactory achievement.

And to be clear, we will never resort to the use of steroids or drugs to "enhance" our training since we are not competing as bodybuilders. We are doing things the natural way: through proper planning regarding both diet and exercise, hard work, and plenty of rest. What about using non-steroid substances to possibly enhance training? I will discuss the potential use of such supplements in Chapter 5.

With the Old Man Muscle workout, we will be training the old school way, using primarily dumbbells. The reason for this is that dumbbells are very effective for not only isolation work (single muscle groups) but also for compound exercises (working multiple muscles and joints at the same time). These are the kinds of exercises that we need to focus on in order to rejuvenate our aging bodies. The Old Man Muscle workout not only has the potential to build significant strength and muscle in the first year, but we get a boost to our falling testosterone levels as well, which helps fight specific effects associated with aging. Dumbbells are easy to manage, and permit full range of motion of joints, working each side of the body independently. This helps prevent imbalances and builds a pleasing, aesthetic physique.

Reasons to train using free weights as dumbbells:	
• Convenience of size, affordability of cost. • Relatively easy to control during the workout. • Enable the establishment of mental engrams.	• Provide a superior range of motion during exercises. • Generate a workout for the core. • Stimulate superior muscle fiber recruitment.

▲ **Table 3.3** Several reasons to work out using "old school" dumbbells as the chosen training modality.

Before embarking on a strength training program, you should be in apparent good health and have obtained medical clearance to exercise. This is because in certain individuals, pre-existing health conditions may be such that a strength training program is contraindicated. Do not begin exercising unless and until your health professional tells you that you are good to go.

WE MUST ADOPT AN AGE-APPROPRIATE TRAINING STRATEGY

Our approach to working out must embody an age-appropriate strategy. We must always keep in mind that when first beginning to train with weights, we must start with light weights in order to learn the proper form. This enables us to establish the correct **mental engrams** (neural networks in the brain that ultimately instruct the muscles and joints how to perform) for each lift, and to avoid getting hurt. After all, we are beginners! Don't worry

– we will increase the poundage, but only gradually over the first year. Because of our age and the fact that our muscles are not yet developed nor accustomed to regular exercise of this type, we must use caution and not over-train at the outset.

Realize too that the muscles and tendons, ligaments and bone, and our joints all will be stressed from each workout. Although this will be a good form of stress, controlled to be within acceptable limits by our program design, it is still stress. Muscles respond differently to stress compared to tendons and ligaments, as do the joints, which do not have the equivalent healing blood supply that muscle has. Joints can become stress injured, requiring a long time to heal, if we lift too heavy, too soon. And at our advanced age, we simply should not lift as heavy as the youngsters, because we have "joint baggage". By this I mean that our joints have already seen many decades of use and possible abuse, meaning wear and tear. Do the words "overuse injury" sound familiar?

According to research, joints do not grow stronger with the heaviest loads the way muscles can – they thin out and become more susceptible to injury. While a young guy seemingly can lift heavy and not worry about his joints (although too much at a young age also translates into problems later on – look at all the former athletes who are in wheel chairs as old men), WE must be cautious.

Moderately heavy work is the eventual goal by the end of year one, rather than lifting excessively heavy weight. We are not in our teens! And we are not professional bodybuilders. We are going for a lean and hard look – it is not typically possible to become big and bulky like the younger guys. We do not have their testosterone and growth hormone levels, or anabolic response to dietary protein. There may be a rare exception to this – and if that happens to be you – congratulations are indeed in order. But for the rest of us, remember: training to become lean and hard is the goal – and we do so through natural effort – not resorting to steroids or other dangerous drugs. So we must train wisely, and not overdo things at first – otherwise the result will be injury, pain and disappointment.

A major goal of ours during the first two years of training is to lose as much excess flab around the belly that we can, while getting some size, definition and vascularization (veins) going on in our arms and overall physiques. What I am getting at here is a full physique makeover. Believe me, this is not easy when you are much over 60, but it is possible. By losing the extra abdominal fat, we will achieve several desirable goals simultaneously: (1) belly fat results in conversion of testosterone to estrogen; the less belly fat we have, the more our dwindling testosterone levels will be preserved and amplify our workout results as well as stall the aging clock; (2) losing belly fat means less overall body fat, which will improve our vascularity; and (3) the less belly fat we have, the more our abs will be visible, and the smaller our waist will appear. We should strive to maximize the difference in the measurement of our waist (smaller) compared to the chest (larger). We cannot do much about the hips and shoulders, although a strong-looking physique will show differences between the shoulders (broadest), chest (broad), waist (narrow), and hips (slightly wider than waist) measurements. There is also the challenge of adding muscle mass while trying to lose weight and fat mass. Often times, folks will lose muscle as they lose weight and fat.

EXERCISE SCIENCE & STRENGTH TRAINING BASICS

To begin with, it should be generally agreed that each of us desires to dedicate ourselves to achieving a higher level of strength, in addition to fitness and health, as we age. Yet, we should appreciate that a person can develop strength and/or fitness but still not be truly healthy. This happens especially if there is underlying disease present (cardiovascular, cancer, diabetes, etc.) or poor lifestyle habits (drug or alcohol abuse, poor nutrition, and so on). However, the more fit a person becomes, such disease issues can be minimized.

Fitness, or physical fitness, is typically associated with (1) a lean body composition with a functional **LBM** (lean body mass), (2) lack of excessive body fat, and a normal **BMI** (body mass index); (3) adequate cardiovascular endurance; and (4) flexibility in terms of all joints being able to move through their entire range of motion (ROM). In persons able to demonstrate all four categories, health issue risk becomes low, and the performing of **ADL**s (activities of daily living) is the expectation. However, the reality is, as we age, we tend to become compromised in one or more of these areas.

Further on the topic of fitness: one can differentiate between what is typically considered cardiorespiratory fitness (or what I like to call "treadmill fitness") as compared to a newer concept, that of metabolic fitness, which is associated with metabolic health. **Metabolic health** is defined as having optimal levels of blood fats (triglycerides), blood sugar (glucose), blood pressure, high-density lipoprotein cholesterol (HDL), and waist circumference. A metabolically healthy individual will exhibit all of these characteristics without the need for any medications. Statistically, only about 10% of Americans are achieving optimal metabolic health. The importance of being metabolically healthy is that such individuals are at lower risk for lifestyle diseases (cancer, cardiovascular disease, diabetes, high blood pressure, obesity). Clearly, a person who is both physically fit and metabolically fit will have the potential to age better and carry on life's daily activities in spite of the passage of many years.

Then too, there is the concept of muscular fitness, which is related to strength training. First of all, **strength** is the ability to create maximal force during a single contraction of a muscle. Powerlifters are focused on achieving massive amounts of strength in order to lift very heavy barbells. Size relates to muscle volume and appearance, and we all know that bodybuilders train to achieve this. The increases in the size of muscle due to training is called hypertrophy. There are actually a couple of kinds of hypertrophy. One is true hypertrophy due to myofibrillar expansion through the action of MPS and satellite cell activity, occurring between workouts. The second kind is **sarcoplasm**ic hypertrophy which is transient – also known as the "muscle pump". It is caused by fluid that accumulates in the sarcoplasm (the area between myofibrils) during the workout but soon drains out. It is believed that this extra fluid temporarily creates increased pressure in muscle tissue and this provides growth stimulus. Turning next to **endurance** – this refers to one's ability to maintain a specific muscular activity over time, such as jogging or cycling.

SKELETAL MUSCLE

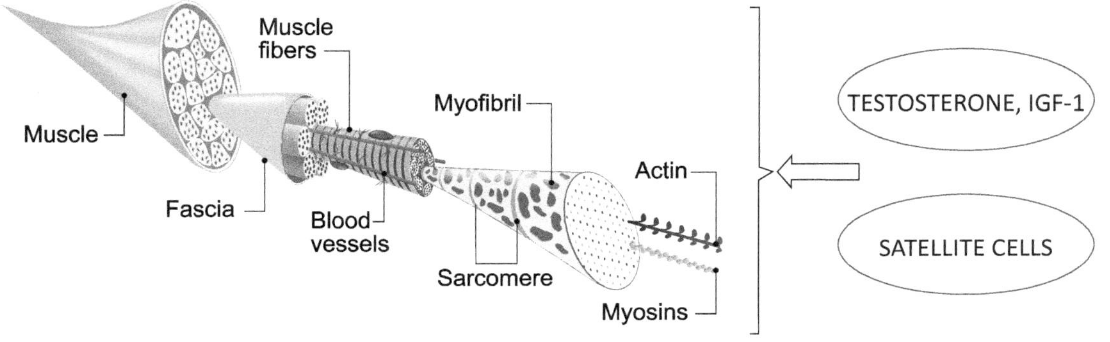

▲ **Figure 3.2** Coming to terms with the mechanism of hypertrophy: myofibers or myocytes – individual muscle cells – increase in size following a workout due to proliferation of actin and myosin myofilaments within myofibrils (especially the fast twitch type II), as aided by testosterone hormone. In addition, muscle stem cells called satellite cells respond to the stress of mechanical muscle damage from training by fusing together and to muscle fibers, which leads to muscle repair and growth. This is aided by the hormone IGF-1.

I want to point out that in strength training, it is not only the body's muscles that get a workout. A key fact to realize is that none of the muscle contractions that are so essential in strength training would occur without the involvement of the nervous system. **Neuromuscular control**, which is the communication between the brain and muscles, is a critical relationship that influences all variables of muscular fitness. So when we train, we are training the nervous system as well as the muscular system. And beyond that – we need to train in a manner that facilitates an effective and efficient inter-relationship between the two systems. This brings us to the topic of muscle fibers, motor units, and mental engrams next, and what is called the mind:muscle connection toward the end of the chapter.

Types of Muscle Fibers and Motor Units

As we all are probably aware, the muscles that we will be developing through the Old Man Muscle workout are termed skeletal muscles. We will be focusing on developing strength in the skeletal muscle in our arms, legs, chest, back, shoulders, and core. Each skeletal muscle is an organ that consists of various integrated tissues, including the skeletal muscle cells themselves, blood vessels, nerve fibers (neurons), as well as connective tissue. Due to their long, cylindrical shape, skeletal muscle cells are commonly referred to as **muscle fibers**.

Muscle fibers are the building blocks of our muscles. They are classified as fast twitch or slow twitch, and whether they are primarily oxidative or glycolytic. Looking at the muscle fiber table provided, high oxidative (associated with **slow twitch** "red" fibers) means the oxygen capacity of muscles is increased, which is good for building stamina but not power. It takes these muscles a longer time to fatigue (think marathon runner). High glycolytic (associated with **fast twitch** "white" fibers) means energy and force production of such muscles is increased, which is good for building muscle and power (think sprinter). Such muscle has more strength to overcome resistance forces.

TYPE I	TYPE IIa	TYPE IIx
High Oxidative Low Glycolytic SLOW TWITCH ENDURANCE	Med Oxidative Med Glycolytic MED TWITCH POWER	Low Oxidative High G lycolytic FAST TWITCH POWER

▲ **Table 3.4** Muscle fiber type classification. Muscle fiber types are linked to their metabolism pathways in that they will be aerobic or fat-burning for endurance (Type I fibers), or anaerobic and glucose burning for power (Type II fibers).

Sadly, as we age, we lose fast twitch fibers, and more of them convert to slow twitch. What this essentially means is that if we do not strength train, our muscles lose their explosive power and our bodies become weaker. Research is investigating the regaining of fast twitch fibers through strength training in old age, as well as the possibility to convert from slow fiber type to fast. Related to this, later in this chapter, look for a discussion centering on what is termed BDNF when I discuss adaptations to exercise. By the way, the terms "slow" and "fast" have to do with the conduction velocity of nerve impulses from neurons to the muscle fibers (we discuss what a motor unit is next), and therefore, to the speed of muscle contraction. Those words also align with the rate at which ATP is generated in muscle either by the oxidation of fat (slow) or glucose (fast).

We already learned that muscle fibers are integrated with neurons (nerve cells) in motor units. This is very important, because as we train, we are establishing and developing new connections between the brain and its neural pathway to our body muscles (neurogenesis). Our muscles cannot contract without neurons telling them to do so. So as we train, we create this communication pathway.

When we train with weights using the Old Man Muscle workout, the CNS (central nervous system) sends nerve impulses in order to communicate with and activate specific motor units and associated muscle fibers in what are called **prime mover** (agonist) muscles. As muscles work during our lifts, muscle fibers are contracting in order to generate forces to move the weights. They utilize various fuel sources, especially glucose and creatine phosphate, in a process called anaerobic energy ATP production. **Anaerobic energy**, by definition, generates ATP without the need for oxygen to be present in the cell environment, and can do so very rapidly with a reduced number of enzymatic steps required. The fact that we can rely on an energy process that doesn't require oxygen is a good thing – because we would be unable to breathe in enough oxygen through our lungs for this type of energy need.

On the other hand, when fat (instead of glucose) is oxidized for ATP energy production, the process involves oxygen (is aerobic), plus many more enzymatic steps in the body are needed, which means it takes longer to generate these ATPs. Going along with the notion of a marathon runner versus a sprinter, the body selects the manner in which ATP is generated accordingly. In order to execute the burst of power for a quick sprint, the body shifts energy metabolism to the anaerobic breakdown and oxidation of glucose to generate ATP, while for the less intense but long duration demands of a marathon, energy metabolism shifts to fat oxidation in the presence of oxygen. In the aftermath of ATP production, metabolic waste products are generated, including H+ (hydrogen ions), CO2 (carbon dioxide) and lactic acid.

While a discussion of fuel sources for muscle is interesting and important, allow me to take a step back. The way to understand the physical linkage between muscle and nervous system is by understanding what a motor unit is. A **motor unit** consists of a single nerve cell or neuron that innervates (is connected to) a group of skeletal muscles. Motor units therefore are connections between the central nervous system and individual muscle fibers. Neurons receive signals from the brain and stimulate all the muscle fibers of a particular motor unit. As soon as a motor unit receives a signal from the brain, all of the muscle fibers in that unit contract at the same time with full force.

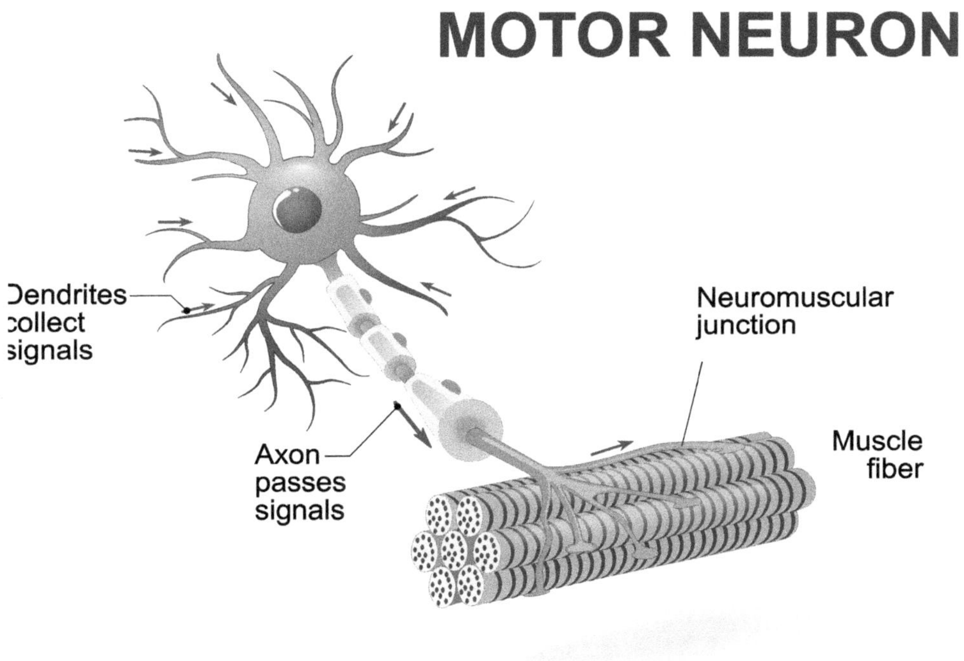

▲ **Figure 3.3** The motor neuron conducts nerve impulses from the CNS to skeletal muscle fibers for contraction and development purposes. A motor unit comprises a motor neuron and all the associated muscle fibers that it stimulates.

The neuromuscular system goes through a cycle when developing strength – that is, specific strength training exercises, as they are first being learned, teach the brain to fire specific muscles, which contract with a new movement. The learning part involves memorized instructions that are created through repeated practice of proper form and execution of each exercise. These mental engrams are specific and establish particular neural pathways (nerve connections between the brain and the muscles) for each exercise, so that we become more efficient when working our muscles. This is why proper form must be practiced – you don't want to establish the firing of the wrong muscles and create an incorrect mental engram.

After many times of practice, mental engrams become established and we don't have to be methodical about what we are doing – the exercise movement becomes automatic. This does not mean that we should not focus on the job at hand, nor not be thinking about each move before we attack the weights – we should! But that has more to do with motivation and ultimate execution than learning the exercise. And so, we can gradually add resistance (increase the weight amount) in order to recruit more muscle fibers to oppose the resistance, to build strength, and adapt to the resistance so we can do more. The progression of increasing the load, recruiting more fibers, gaining strength, and causing neuromuscular adaptation results in the ability to add more resistance as workouts progress, because the new, stronger body has adapted and will continue to adapt by requiring more load (resistance) in order to keep growing and strengthening. Building strength and muscle are part of the overall response called the **training effect**, and it is something all individuals engaged in strength training look forward to experiencing. The training effect involves a whole series of what are called adaptations, which include brain, cardiorespiratory/cardiovascular, endocrine and other metabolic as well as neuromuscular changes – all for the better! This is what gives rise to the concept of exercise as medicine.

To restate, an engram is a stored motor pattern that allows us to be neurologically efficient. It is a set of instructions stored in the brain that results from performing a particular activity (or exercise) correctly many times. Neurological efficiency occurs when a stored motor pattern (engram) is focused on the exercise and nothing else. This is when power, strength, and size will be gained. When instead we perform an exercise incorrectly and are not focused, we are neurologically inefficient, wasting time and effort. When this happens, we need to stop and re-learn the motor pattern. Interestingly, while aerobic exercise classes are conducted often to the beat and sound of music, the kind of strength training we engage in will be negatively impacted because the mind will be focusing in part on the music when it should be totally focused on proper execution of each exercise.

Training for Hypertrophy

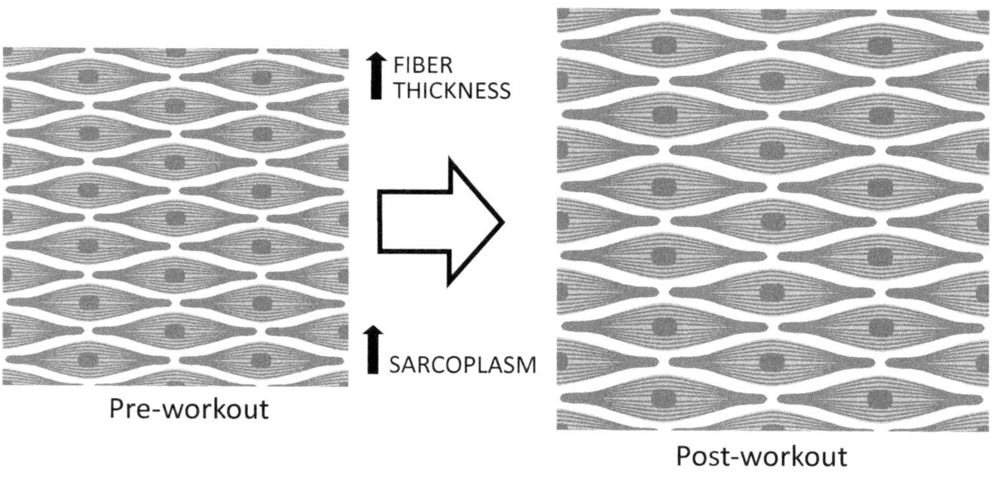

▲ **Figure 3.4** Muscle anatomy at the muscle fiber level reveals two components that contribute to hypertrophy. Growth is thought to occur by increasing either muscle fiber thickness or increasing the volume of fluid in the sarcoplasm of muscle cells. The sarcoplasm in muscle cells serves a similar function as the cytoplasm in other body cells. There is storage of glycogen (within glycosomes) and the presence of myoglobin in the sarcoplasm of muscle fibers.

This is an important issue, and I'm certain to talk about it again in the next chapter. For now, let's appreciate that as older men, we are mostly performing strength training to develop strength, but we will also build muscle size. However, we may not be heavily focused on achieving the maximum size (hypertrophy) of our muscles the way a body builder or competitive athlete might. Even so, there is nothing wrong with wanting to have bigger arms, shoulders, chest and so forth – to change one's physique in other words. So it is good if I use some page space to discuss training for muscle size a bit before moving on to the next topic.

First of all, the research on training approaches that result in the most muscle growth is somewhat conflicted. The two schools of thought on it are (1) to train with heavy weight and low repetitions, and (2) to train with less weight but higher reps.

TRAINING GOAL	#SETS	# REPETITIONS	REST INTERVAL	INTENSITY
Muscular endurance	2-3	>12	<30 seconds	60-70% 1 RM
Muscular hypertrophy	3-6	6-12	30-90 seconds	70-80% 1 RM
Muscular strength	2-6	<6	2-5 minutes	80-90% 1 RM

▲ **Table 3.5** Examples of recommendations for training muscle. While these figures are for comparison purposes, based upon a publication by Bachle and Earle (2008), experts are not agreed on these specifics, and if you perform an internet search, you will find diverging opinions.

Training volume is thought by many to be the key. Training volume is simply this:

Training VOLUME = (lb. amount of dumbbell lifted in an exercise) x (# of reps) x (# of sets)

For example, let's imagine a very simple workout routine of three exercises, the bench press, squat, and curl, each performed with 20-lb. dumbbells, for two sets of 10 reps. Suppose we are told that for each exercise, one 20-lb. dumbbell must be held in each hand (the factor of "2" in the calculations below). Then, we can proceed to calculate the training volume for each exercise and sum them in order to calculate the training volume of the whole workout. It would be determined like this:

- *bench press 20 lb. x 2 = 40 lbs. lifted each rep, x 10 reps, x 2 sets = 800 lbs.*
- *squat 20 lb. x 2 = 40 lbs. lifted each rep, x 10 reps, x 2 sets = 800 lbs.*
- *curl 20 lb. x 2 = 40 lbs. lifted each rep, x 10 reps, x 2 sets = 800 lbs*

Total training volume for this workout: 800 + 800 + 800 = 2,400 lbs. lifted.

This gives the volume in amount of weight lifted in pounds per exercise, which is then summed for all the exercises in the entire workout to yield total training volume for the workout. Since it is believed that training volume is essential to muscle development, strength, and hypertrophy, and since progressive training demands that we increase overload and training volume over time, we can track the actual training volumes of our workouts to be sure that we do not remain at the same level and fail to make gains, as well as plan them in advance.

The issue regarding the superiority of high reps/lower weight or high weight/fewer reps for increasing muscle size I do not believe has been firmly settled in the research literature. On top of that, it makes a difference if studies trying to evaluate training effects are conducted in men, women, untrained individuals or trained, and what age they are. There are studies that focus on men past sixty, fortunately – but more work is needed to fully understand the facts. Lastly, we each have our individual differences (genetic, physiological, metabolic, emotional, and otherwise) and respond differently to the same resistance training exercises. This is because each one of us will have our own unique adaptive response to training that will impact our ability to add mass to a greater or lesser degree. I hasten to add that too much stress in one's life, a poor diet, or lack of sleep can significantly reduce one's ability to grow muscle. Why? These factors can impact the endocrine (hormonal) growth factors that are part of the muscle growth equation.

I want to discuss three areas relevant to this issue: time under tension (TUT), mechanical overload, and metabolic overload. Superimposed upon these is the overload principle that every student of exercise science learns. The **overload principle** states that in order to stimulate muscle growth, it is necessary to apply a physical stimulus at an intensity greater than the body is accustomed to receiving. As we all know, that is the reason to lift weights such as dumbbells, which act as physical resistance. This is why weight training is a form of resistance as well as strength training.

What is the difference between mechanical overload (or stress) and metabolic overload? **Mechanical overload** describes the structural damage to muscle fibers as the result of lifting weights. In order to achieve this, we go back to the overload principle. The muscles must be subjected to a stress that exceeds what they typically experience. Mechanical overload specifically represents a stress applied to the physical structures of the motor units. But the molecular and cellular events that follow a bout of exercise are quite complex and not fully understood.

What about time under tension? In order to understand it, let's go back to thinking about motor units a bit. Muscles contract in response to signals from the muscle motor units, so it is important to activate as many of them as possible when exercising if the goal is to become stronger and also bigger. The more a muscle is contracted during exercise, the more motor units become activated. The activation of motor units is enhanced by the amount of time - the duration – that a muscle stays contracted. This state of contraction that can be achieved by the person who is lifting is called the **time under tension**.

What we are adding here is that the manner of training using time under tension may augment muscle hypertrophy. There are actually several considerations: (1) how much weight (the resistance load) is being lifted during each rep; (2) how quickly or slowly the muscle is being contracted (the pace or tempo of lifting); (3) the total number of repetitions performed; and (4) performing repetitions to fatigue or failure (the intense burning sensation that causes the person to stop lifting).

Regarding point (4), when an exercise is performed to the point of fatigue, as evidenced by the inability to complete another repetition, this generally indicates that all muscle motor units have been activated. In turn, motor units can activate both type I and type II muscle fibers, in sequence. First activated when we start an exercise are type I, and if we proceed and develop sufficient intensity and/or volume, type II. For muscle size to increase, stimulation of type II muscle fibers is critical. Therefore, exercises that are known to activate type II fibers are traditionally utilized in weight training for this purpose.

On the topic of **metabolic overload**, we must recall that ATP energy drives muscle contraction. Importantly, the source of ATP during power activities like lifting weights is the oxidation of glucose anaerobically in muscle cells. Where does the glucose come from? Muscle has the ability to store water and also glycogen. Glycogen is a large molecule (a polymer) that is made up of many glucose units. When muscle contracts, it calls upon the glucose stored within glycogen in order to make ATP so that the muscle can keep on contracting. At some point, if the exercise is continued to fatigue, glycogen (hence glucose) will be used up. This process of lifting enough times (high reps) in order to reach fatigue results in depleted muscle glycogen and the production of lactic acid/lactate, and hydrogen ions (H+). Hydrogen ions create a decrease in blood pH, which is toward the acid range. This transient shift in blood pH is believed to help stimulate the release of growth hormone. During recovery, muscle tissue will regain glycogen plus water, and with more training, the person becomes more efficient in this type of overload process, which contributes to post-workout hypertrophy.

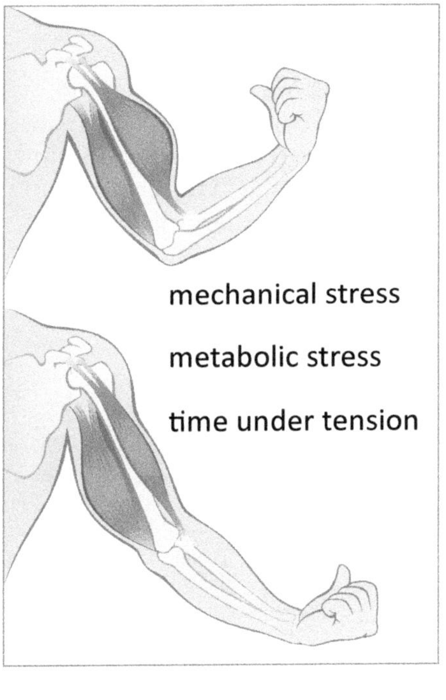

▲ **Figure 3.5** Muscle hypertrophy is the result of exposure of muscle to stress. The kinds of stresses that can make muscles grow are (1) mechanical stress, (2) metabolic stress, and (3) time under tension. Researchers currently believe that mechanical stress is the primary stimulus, involving satellite cells and muscle remodeling post-exercise, although the other stresses play a role as well.

In training, after the introductory settling-in period of two-three weeks for most untrained older men, a slow progression of mechanical overload and/or metabolic stress can be incorporated into the workout plan. Each individual will respond his own way and perhaps better to one form (mechanical vs. metabolic vs. combined or alternating) over another. One way in which to identify that our training is effective is to see the evidence: the growth of muscle. However, visible muscle growth takes quite a bit of time, and more so for older men. In my case, I could feel the muscles I was training after only 2-3 weeks, but I could not really see them growing at such an early stage. And since you view yourself in the mirror on a daily basis, it is difficult to notice small, incremental changes in one's physique. What is a better indicator is when someone else notices and makes a comment, or you fit into your clothes differently. Pants become loose around the waist, shirts become tighter in the upper arm and chest! Be prepared to wait months and more for such changes to occur. Remember, it takes years for bodybuilders (those not taking steroids) to achieve big and ripped physiques, and these are men in their twenties with naturally high hormone levels! For us, patience and persistence are essential.

Satellite Cells

I feel the need to discuss a bit about **satellite cells** at this point. These cells are physically located in the vicinity of muscle fibers (myofibers) and are termed "satellite" for this reason. Our skeletal muscle has the remarkable ability to repair itself after workout-induced trauma, which is termed "micro-trauma" with respect to motor units and associated muscle fibers. Satellite cells are quietly sitting around waiting to spring into action to repair and build muscle fibers when needed. Such regeneration is a highly orchestrated process involving the activation of various cellular and molecular events, and specifically involves satellite cells. These cells actually proliferate, differentiate, fuse, and lead to new myofiber formation plus the reconstitution of functional contractile proteins (**actin and myosin** myofibrils composing **sarcomeres**, the basic functional unit of muscle). Satellite cells are able to donate their nuclei (myonuclei) to myofibers during repair, not once but in multiple cycles. Scientists would summarize this phenomenon by saying that "progressive myonuclear accretion leads to functional muscle hypertrophy."

The way that satellite cells work is that once activated from the quiescent (not metabolically active) state, they produce what are called **MRFs** (myogenic regulatory factors). MRFs function as **transcription** factors, which means they function to make certain that the correct genes are expressed in specific cells of the body at the right time. In this case, the transcription factors make sure that the quiescent satellite cells undergo sequential and complete differentiation into new myofibers. The specific MRFs are MyoD, Myf5, Mrf4, and myogenin. These factors act on the quiescent satellite cells and transform them as they differentiate into mature myofibers. This overall process is called **myogenesis**, the formation of new muscle. But there is much more to the story. Recall for instance mTOR, the key regulator of protein synthesis. Well, it turns out that satellite cells (like other cells in the body) possess mTOR, which apparently helps express myogenic genes in satellite cells in order for them to properly function.

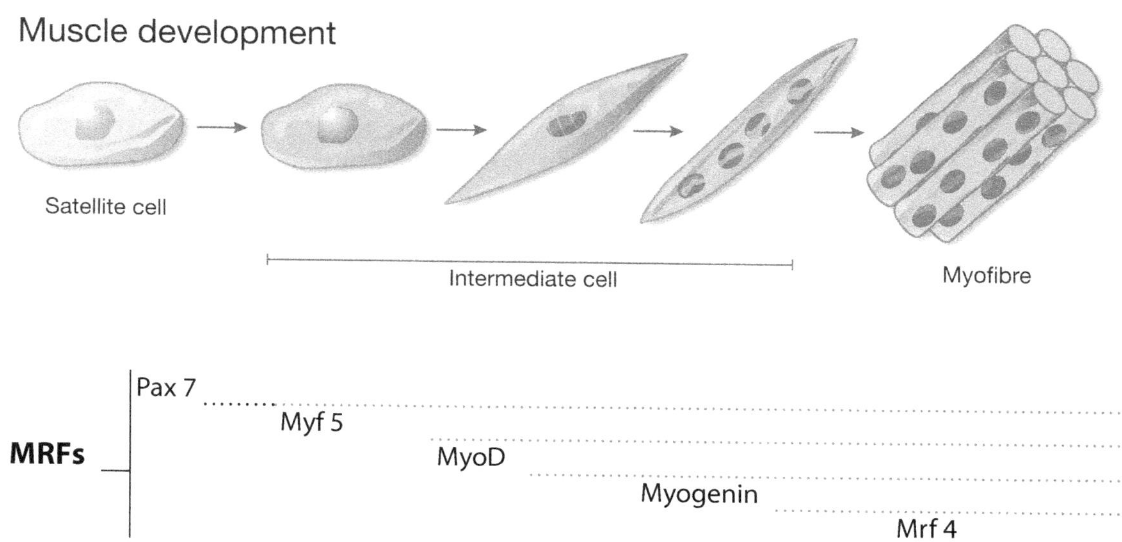

▲ **Figure 3.6** The process of myogenesis, the formation of new muscle cells (myofibers), which happens in response to muscle damage created by strength training workouts. This is the reason for new muscle growth after we work out. As depicted, satellite cells are at the heart of the process. What is shown is not only the differentiation of immature quiescent satellite cells into mature myofibers, but also the expression of the various MRFs at different stages of myogenesis. For example, Myf5 is expressed earliest while Mrf4 is last. As the various MRFs are produced, this is called "upregulation". There is another transcription factor not shown, Pax7, which is expressed in the quiescent cells, but becomes downregulated as the cells proliferate. When the differentiation process reaches completion (far right), the mature terminally differentiated cells may either fuse with pre-existing myofibers, or fuse together to create a totally new myofiber. The end result is greater muscle mass, one cell at a time. Notice also that muscle cells are multinucleated – this is rare for cells, but allows for greater cell size. As new muscle cell nuclei (myonuclei) are recruited by overload training by the activation of satellite cells, this contributes to hypertrophy.

By the way – we all have likely heard of the disease called muscular dystrophy (MD). Well, much study regarding molecular mechanisms in MD has led to increased understanding of what happens when we lose muscle mass and function. Specific to satellite cell activity, there is the hope that therapies can be developed. A case in point is regarding dystrophic mice, a lab animal often chosen in order to study disease mechanisms. Did you know that the exhaustion and depletion of satellite cells in such mice was reversed through the application of **nitric oxide** (NO)? And did you know that NO itself is a potent signaling molecule and stimulator of muscle mass, function, and regeneration? Yes, studies have shown this. However, clinical trials using NO therapy in humans have produced only mixed results.

That's enough regarding dystrophic mice, MRFs, and satellite cells. Admittedly, this has been a challenging load of technical information for us to make sense of. As we see, the mechanism of muscle growth is highly complex, and involves much more than knowing about amino acids, protein, and MPS (muscle protein synthesis).

One last point, which could have been mentioned in the chapter that discussed the challenges we face as older men. It should be remembered that even if we do not work out, the body is turning over muscle tissue daily – breaking it down, and rebuilding it. If we do not strength train, we will not sustain muscle during this process, and we even start to lose mass due to **sarcopenia**. Well – you guessed it – as a result, as we age, muscle regeneration via satellite cells cannot compensate for the loss of muscle tissue. Specifically, the depletion of the satellite cell pool, and/or the loss of satellite cell functionality, are age-related realities thought to be factors that cause our inability to maintain muscle mass as we age (unless we take appropriate steps, that is).

Training Intensity

Training intensity has already been mentioned within the context of hypertrophy. When intensity is stated as 70%, this actually refers to a concept known as %RM, for the percent of repetition maximum. A 70% of 1-RM would be a weight lifted that is 70% of the maximum weight that the person could lift only one time in a given exercise. This means the same thing as a one repetition maximum, and is typically designated **1-RM**. For example, if you can only bench press 100 lbs. one time and not successfully perform any further reps, then 100 lbs. is your 1-RM, and 70% of your 1-RM for this exercise would be 70 percent of 100, or 70 lbs. This is a good training intensity for both muscular strength and hypertrophy, according to many reports.

For an older person, it is wise to approach that training level gradually, however. Also notice – since lower load weights enable you to lift more repetitions, you might be able to complete 10 repetitions at 70% of 1-RM, and 15 repetitions at 60% of 1-RM. The connection between %RM and training intensity is that the higher the percentage of the 1-RM, the greater the training intensity. It has been suggested that lower training intensities (e.g. less than 60% 1-RM) may not build muscle and strength as effectively as higher intensities, but are safer in terms of less risk of strain or injury on joints and soft tissue. However, there is some recent research that throws these assumptions into question, asserting that **training volume** for a range of % 1-RM values may be the best predictor of muscle and strength development. But here again, we should always keep in mind that as older men, we do not have the same ability to recover from repeated high training volume and intensity workouts, and may perhaps even suffer adversely in the long term. Therefore, if our desire is to train indefinitely into advanced old age, we must decide which way to go with intensity.

ADAPTATIONS

I have already used the word "adapt" in this chapter, did you notice? In this section, we will focus on several exercise-specific adaptations. This means the good changes that the body experiences when it is challenged by exercise. One of the amazing things about the human body is its ability to respond to exercise in a multitude of positive ways. Such an ability to change is termed plasticity. And for us as older men, it is equally amazing and gratifying to know that the body will act this way even at an advanced age. So, what we will next consider is that the body's response to exercise manifests in various, specific ways, called **adaptations**. Such adaptations are normal and must be understood in order to plan successful long-term training strategies. We will briefly review the following adaptations: cardiorespiratory and cardiovascular systems, **endocrine system**, muscle, and nervous system including the brain.

ADAPTATION BENEFITS		
DISEASE PREVENTION		LESS INJURY RISK
	INCREASED PHYSICAL CAPACITY	
IMPROVED METABOLIC FUNCTION		HEALTHIER BODY COMPOSITION

▲ **Table 3.6** There are numerous adaptation benefits to physical activity in general, and strength (resistance) training in particular. Within these categories, we could identify additional details, such as improve overall strength, flexibility and mobility – as well as overall functional capacity.

Before we examine each of these, it should be mentioned that it has been observed that a single bout of exercise results in an **acute systemic response**, whereas chronic adaptations occur when exercise is habitual over the long term. Thus, when we first start training, it results in an acute response that develops into desired chronic adaptations if we stick to regular training. If we do, we will achieve the training effect mentioned earlier. This is our goal – to make training a permanent lifestyle change so that the beneficial adaptations that we will experience will remain with us as long as possible.

So many of the adaptations that occur when muscle tissue is stressed through a workout are the result of cell-to-cell communication. This basic ability - and activity - of cells, is as central to exercise adaptations as it is to energy metabolism. The latter, which takes into account the effects of food nutrients on the diversity of cell functioning, specialized tissue growth and functioning (like muscle), as well as whole body effects, is highly complex. It involves knowing about signaling molecules and the pathways they follow, the binding of transcription factors to cell receptors in order for genes to become activated and produce products that cause targeted effects, upregulation and downregulation of such gene-produced cell products, upstream and downstream events in signaling pathways, and more.

If we were to begin a study of the impacts of just two different molecules, **AMPK** (AMP-activated protein kinase) and mTOR (mechanistic target of rapamycin), and their interactions and impacts on cellular metabolism specific to protein synthesis, it would take a book in order to do so. We would have to understand the importance of what is called phosphorylation of important regulatory molecules via protein kinase enzymes, which can both activate or deactivate such key molecules. It would be critical to investigate specifically how AMPK and mTOR control anabolism and catabolism by switching on and off (hence, regulating) a limited set of master metabolic switches. These two molecules influence health and disease, not merely protein synthesis and cell growth.

PHYSIOLOGICAL ADAPTATIONS	**CARDIOVASCULAR SYSTEM** Improved blood pressure Stronger cardiac muscle **CARDIORESPIRATORY SYSTEM** Improved ventilation **NEUROLOGICAL SYSTEM** Positive brain effects Efficient motor learning **MUSCULOSKELETAL SYSTEM** Muscle hypertrophy Muscle contractile strength Increased tendon strength Increased ligament strength Increased bone density **ENDOCRINE SYSTEM** Increases in anabolic hormones Improved insulin sensitivity

▲ **Table 3.7** Five categories of body adaptations that are related to the adaptation benefits mentioned.

In general, AMPK inhibits anabolic processes (which is actually a good thing when a person is starving) while promoting catabolic processes, while mTOR does the opposite. AMPK actually inhibits mTOR activity (as mTORC1 complex). What does all of this have to do with training to build muscle? Well, directly related to achieving muscle hypertrophy through exercise and protein and specific amino acid intake, mTOR activation is required in order for both myofibrillar muscle protein synthesis and skeletal muscle hypertrophy to happen. This occurs when we strength train and also consume adequate protein, which together "overcome" any inhibition caused by AMPK (which inhibits skeletal muscle hypertrophy).

Well, that was a mouthful. I want to leave this molecular discussion right there - and return to the notion of adaptations. What specifically is known about them, which body systems are involved, and how beneficial are the adaptations anyway?

Cardiorespiratory and Cardiovascular System Adaptations

Due to the demands of exercise, the circulatory and respiratory systems are able to adjust their functioning in order to, among other things, meet oxygen intake and delivery needs of cells, and maintain acid-base balance. In fact there is a hidden connection between muscles and respiratory efficiency, which will first be addressed. Adaptation to exercise occurs in the mitochondria of muscle cells. Mitochondria are small organelles found in these cells, and are responsible for generating power at the cellular level. The source of this power is the **ATP (adenosine triphosphate)** molecule spoken of previously. The type of training – aerobic (think treadmill) or anaerobic (think strength training), and their predominant muscle fiber types determines the abundance and activity of these organelles. Mitochondria are important in aerobic metabolism and are prevalent in the type I (slow twitch) muscle fibers, while being limited in fast twitch fibers. With aerobic training (which also can be somewhat approximated in anaerobic strength training when rest intervals and the pace and duration of the workout is controlled), there is the potential to expand the mitochondrial output which results in greater oxygen uptake and utilization capacity, which is called aerobic power or **VO2 max**. Defined as the maximal volume of oxygen that a person can use per minute per kilogram of body weight, the VO2 max is considered the best indicator of cardiovascular endurance.

Training Intensity and Ventilation

Within the context of exercise or training intensity, **ventilation**, as most readers are aware, has to do with our breathing. We inhale (inspiration of air with oxygen) and exhale (expiration of carbon dioxide) in a rhythmic pattern. As we engage in strength training workouts, we will learn the importance of establishing a breathing pattern for each exercise. The tempo of our ventilation changes (e.g. increases) based upon the intensity of the exercise. Ventilation increases as we breathe more deeply (volume of air) and more frequently (rate of breathing) based upon exercise intensity. When we work out, this is one of the most noticeable things – in addition to an increase in heart rate, and sweating.

Scientists discuss **ventilatory threshold** and VO2 max to explain what is happening to our breathing during exercise. It turns out that we go through a couple of stages called VT1 and VT2 as we progress from a state of rest to light and then more intense exercise. Breathing and the ability to carry on a conversation ("the talk test") are markers indicating whether the exerciser has achieved VT1 or VT2, or their VO2 max. This all goes back to the adaptive response of the cardiorespiratory system to exercise. As we become more trained, our ventilation efficiency (as VT1 and VT2) improves, as does our VO2 max. We can exercise in VT1 and VT2 for longer periods, although this is more applicable to endurance training rather than strength training, because of the latter's brevity, and the rest periods between sets and exercises. Strength training is not marathon running. However, a more highly strength trained person should be able to manage intensity better compared to an untrained person.

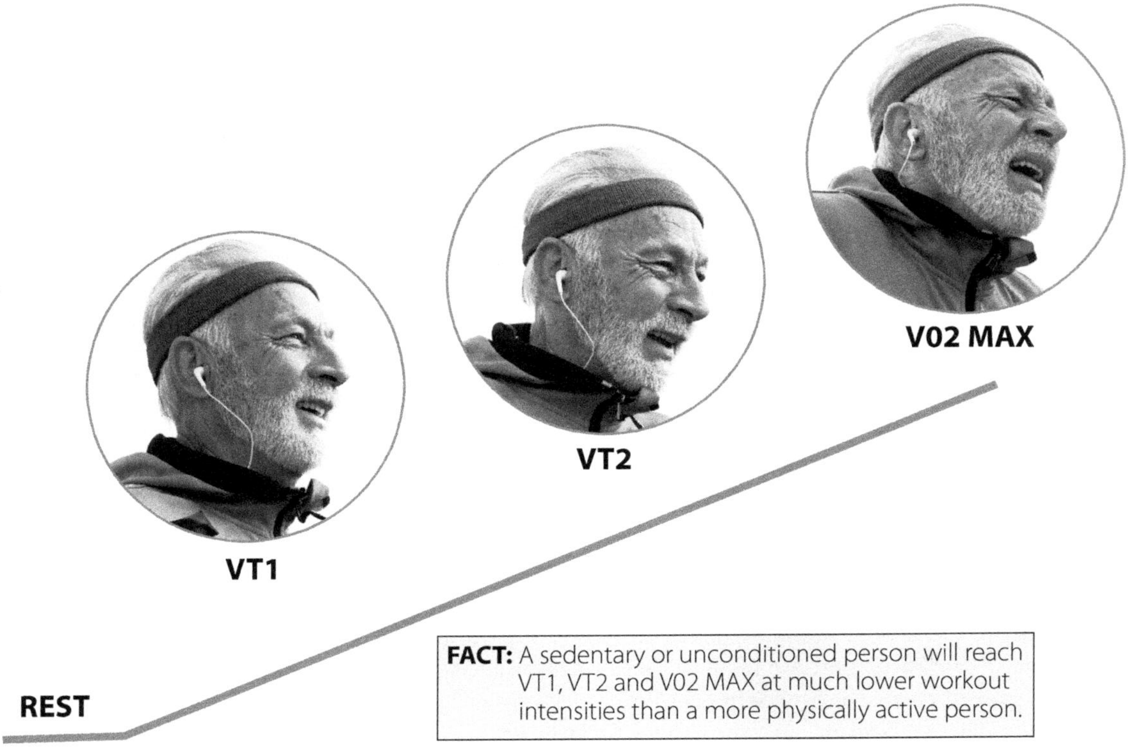

FACT: A sedentary or unconditioned person will reach VT1, VT2 and V02 MAX at much lower workout intensities than a more physically active person.

▲ **Figure 3.7** After a resting person begins to exercise, he will reach VT1, the first ventilatory threshold. A person who is at VT1 finds speech challenging while exercising. Continued intense exercise brings the person to VT2 (also known as the lactate threshold). Lactic acid (as lactate) has accumulated in the blood, and the person needs to breathe so heavily that he can no longer speak. VO2 max corresponds to the maximal consumption of oxygen and the absolute maximum exercise intensity that he can achieve. The person at VO2 MAX will become totally exhausted and fatigued, unable to continue exercising.

Much of the research on strength training in older adults has targeted bone health, physical function and quality of life in older adults. I had mentioned in Chapter 2 that due to strength training there can be an increase not only in cardiac output or blood flow, but also temporarily in heart size and **blood pressure** during exercise.

Training Intensity and the Heart Rate

As we train, it would be a good idea to keep track of our (a) heart rate, and (b) blood pressure. These values serve as excellent indicators of how hard we are working – whether it be too little, too much, or just right. Knowing our heart rate (or pulse) can help us track our health and fitness level. With new advances in sensor technologies, we are now able to track our heart rate and blood pressure using chest band and wrist strap fitness trackers. Although originally developed for those who participate in running and endurance type activities, they can also be useful for our purposes. Such devices will provide quick information on the effects of a workout exercise-to-exercise, as well as comparing such data over time. We would like to see a decrease in our blood pressure, and also in our heart rate, as adaptations due to training. The "Old School" way to determine heart rate is to take your pulse on the inside of your wrist, on the thumb side. Use the tips of your first two fingers (not your thumb) and press lightly over the artery. Count your pulse beats for 10 seconds and multiply by 6 to find your beats per minute (bpm).

Resting heart rate: The resting heart rate is the number of times the heart beats per minute at rest. For most people, between 60 and 100 beats per minute (bpm) is normal. The rate can be affected by factors like age, stress, anxiety, and how physically active we are. Studies have found that a higher resting heart rate is linked with lower physical fitness and higher blood pressure and body weight. So, when it comes to resting heart rate, lower is better. It usually means your heart muscle is in better condition and doesn't have to work as hard to maintain a steady beat.

Target heart rate: The target heart rate is the recommended or desirable heart rate to achieve during exercise. Metabolic benefits are achieved when we work out at the target heart rate. The target heart rate goal decreases as we age, year by year. For example, at age fifty, the target heart rate for an untrained beginner when exercising is between 104 and 128 bpm. For a sixty-five-year-old beginner, it decreases between 98 and 120 bpm. However, we can see improvements when we become fit, including a more youthful heart rate during exercise, along with the concomitant metabolic benefits (lower blood glucose, weight loss, lower blood lipids, and lower blood pressure). This is because with training comes improved cardiovascular fitness as well as other adaptations. For example, the target heart rate at the intermediate level (let's say that means 2-3 years of training at minimum) increases significantly, equivalent to that of a younger person. Even in quite advanced age (90+ years), as we become more fit through our workouts, we can expect improvements in this key cardiovascular marker.

TABLE OF TARGET HEART RATES		
AGE	BEGINNER (bpm)	INTERMED (bpm)
50	103.8-128.0	129.8-145.3
55	101.7-15.4	127.1-142.4
60	99.6-122.8	124.5-139.4
65	97.5-120.3	121.9-136.5
70	95.4-117.7	119.3-133.6
75	93.3-115.1	116.6-130.6
80	91.2-112.5	114.0-127.7
85	89.1-109.9	111.4-124.7
90	87.0-107.3	108.8-121.8

▲ **Table 3.7** Target heart rates in beats per minute (bpm) for older men at the beginner and intermediate levels (source: acefitness.org). It is amazing that a person 90 years old in the intermediate category (perhaps 2-3 years of training) can aim for a target heart rate similar to that of a beginner who is 25 years younger! Said another way, with training, a 90-year old can reverse age 25 years in terms of this vital cardiovascular marker!

Blood pressure: What about exercise can impact blood pressure? Changes to the lining of blood vessels can occur that decrease high blood pressure (**hypertension**). Investigations have demonstrated that strength training programs that successfully reduced blood pressure consisted of 7–10 exercises working the major muscles of the body, 3 days per week, at 40%–80% of 1-RM. This is exactly in line with the Old Man Muscle workout. Over time, participants progressed to 2–3 sets of 8–15 repetitions, with 60–180 seconds of rest between sets.

The two types of blood pressure readings are systolic and diastolic. Systolic is named after the phase when the heart beats, called systole. Diastolic is named for the diastole phase, when the heart is in a relaxed state. Both types are measured in millimeters of mercury, abbreviated mmHg. Expect your systolic pressure to be higher than your diastolic pressure. According to the Centers for Disease Control and Prevention (CDC), healthy systolic/diastolic pressures, respectively, are 120/80 mmHg or less. Pre-hypertensive values are between these and 140/90 mmHg. Values over this are considered high.

Resistance training – specifically, weightlifting – has the ability to increase as well as decrease blood pressure. For example, blood pressure can be dramatically increased during resistance exercise by holding your breath during lifting. This is often the case with isometric exercises, too. Bodybuilders often hold their breath to help increase the amount of weight they can lift. In fact, blood pressures of 320/250 mmHg have been reported during heavy lower-body lifts. Higher blood pressures during lifting are seen with very heavy weights for a low number of repetitions versus lifting lighter weights for more repetitions. It is for this reason that people who have heart conditions should avoid lifting very heavy weights or performing isometrics.

Exercise can lower resting blood pressure, although studies measuring resting blood pressure as a result of weight training show mixed results. Weightlifting tends to reduce body fat, and reduced body fat often but not always leads to lower blood pressure. However, long-term weightlifting has been shown to lower blood pressure during subsequent weight-training sessions. Blood pressure has been shown to be lower in bodybuilders than in novice lifters at all common relative training intensities. In studies, at least several weeks of training is required to observe such benefits. Long-term weight training also reduces blood pressure versus what it would be otherwise during other non-lifting physical activities, such as running. This is a sign of a stronger cardiovascular system because less pressure is needed to pump oxygenated blood through the body. In fact, if the heart of a dedicated weightlifter is compared to that of a dedicated marathon runner, the weightlifter will have the stronger heart and the lower blood pressure, all else equal. He will also have a more powerful physique.

Most people think of aerobic cardio exercise in order to reduce the risk for cardiovascular disease. However, recent research shows that strength training can offer the same benefits, requiring less than an hour per week to reduce the risk for a heart attack or stroke by 40 to 70 percent. Also noted were reductions in **hypercholesterolemia**, or high cholesterol. Researchers discovered that an increase in training volume (remember: reps x sets x load) improves cholesterol profiles more than an increase in training intensity.

We would like to see the Old Man Muscle trained person improve his heart rate, blood pressure, and decrease his total and LDL cholesterol (if high) and raise HDL cholesterol (if low) by the end of the year compared to when he began training.

Endocrine System Adaptations

Next, a quick word regarding the response of the endocrine system to exercise-induced stress. The endocrine system comprises the hormones of the body, which are peptide (protein molecule) chemical signals that travel from the organ of their origin to target tissue in order to cause some sort of effect. Hormones respond both during the exercise and after, depending upon which particular hormone we are dealing with. Some of the key hormones that are responsive to exercise are cortisol, glucagon, GH (growth hormone), insulin, IGF-1 (insulin growth factor 1), and testosterone. Certain of these hormones induce catabolism (break down molecules and tissue), while others are anabolic (build up or synthesize molecules and tissue).

Generally speaking, the muscle damage incurred through strength training provides a stimulus for hormone secretion. The overall patterns of our daily activities that include any exercise time will impact the balance between anabolic and catabolic hormone secretions and effects. Thus, the manner in which we are physically active is a determinant of hormonal response. When we train with moderate to heavy lifting, we generate sufficiently high levels of mechanical force, damaging muscle protein. This is what stimulates the production of the anabolic hormones which can then promote protein synthesis and muscle growth and repair.

Anabolic hormone adaptations can be acute (secretion following exercise), or long-term. In the acute response, the goal is to repair and build muscle tissue. In the long-term scenario, an increase in the receptor sites and binding proteins for these hormones occurs. We can consider this a potential increase in hormone availability, efficiency and effectiveness.

Cortisol

The hormone **cortisol** is classified as a glucocorticoid from which we can infer that it is involved in glucose metabolism. Glucose is especially needed by cells in times of stress (cortisol is often referred to as the "stress hormone"), and cortisol helps ensure there is enough around. Cortisol is secreted by the adrenal glands that sit atop the kidneys. During extended exercise, times of stress, or low carbohydrate intake, cortisol promotes fat and protein breakdown, which generates keto acids. However, when stress levels are high and sustained, so also will be cortisol levels, and that is when this hormone can become problematic, by affecting our sleep patterns, causing constant fatigue, decreasing our ability to fight off infections, and so forth. Due to its ability to break down fats (triglycerides) in the liver and protein in muscle, cortisol is a catabolic hormone.

GH (or HGH, human growth hormone)

Growth hormone is a major anabolic hormone secreted by the anterior pituitary gland. It facilitates protein synthesis, nutrient transport, and stimulates cell growth. Hormones in general tend to follow circadian rhythms and are secreted at set times during the day or night. Growth hormone peaks during REM sleep, and its production in the body is amplified by high intensity strength training.

Glucagon

Glucagon is a hormone secreted by the alpha cells of the pancreas and acts in opposition to insulin. This means that it senses when blood glucose levels are low (hypoglycemia), causing the breakdown of glycogen (stored in muscle and liver), a process called **glycogenolysis**, to produce free glucose units. Additionally, **lipolysis** (fat or triglyceride breakdown to yield free fatty acids) is stimulated in the liver. Through the process of **gluconeogenesis**, these breakdown products can be assembled into glucose. The glucose is taken up by the circulatory system, restoring blood glucose levels to normal. Due to its mode of action, glucagon is a catabolic hormone.

IGF-1

Insulin-like growth factor (IGF) is similar in structure to insulin and is produced in the liver. IGF-1 supports the function of HGH to repair muscle damaged during exercise, and stimulates the proliferation and differentiation of satellite cells – making it an important anabolic hormone that aids muscle growth.

Insulin

Insulin, like glucagon, is secreted by the pancreas. Insulin is produced by specialized cells called beta cells in response to dietary carbohydrate. Dietary carbohydrate is processed through digestion and absorption into glucose and is transported via the blood circulation to body cells, particularly muscle, liver, and brain. In order to get glucose into these cells, insulin is required. The conversion of blood glucose into stored muscle glycogen happens thanks to this hormone. During exercise, muscle can call upon this stored glucose for anaerobic ATP energy needs. In addition, research has shown that insulin facilitates the activation of the mTOR pathway, important in MPS. Insulin therefore is an anabolic hormone.

Testosterone

The primary male sex hormone and steroid hormone **testosterone** is perhaps the best known anabolic hormone. It is primarily produced by the Leydig cells of the testes in males, with smaller amounts produced by the adrenal glands. Most testosterone is transported bound to a specific plasma protein called sex hormone binding globulin (SHBG). The rest is termed free testosterone, which is the bioactive form. Skeletal muscle growth is made possible largely by the action of this hormone. It also increases bone density. Like growth hormone and IGF-1, testosterone is produced in response to muscle protein damage resulting from strength training. Research indicates that testosterone is involved in the activation of satellite cells and what is called "myonuclear accretion", enhancing muscle protein synthesis and hypertrophy.

Vitamin D (cholecalciferol)

You might be surprised to see a vitamin listed in a section on hormones. Well, back in the late 1980s when I was a graduate student at Virginia Tech in the Department of Nutrition and Foods, one of the first assignments I had was to write a paper answering the question, "Is Vitamin D a vitamin, a hormone, or both?" Well you guessed it – I answered both. However, I would add that yes it is classified as both, but since it can be synthesized in the skin upon exposure to sunlight, technically that disqualifies it from being a true vitamin! The International Union of Pure and Applied Chemistry's Commission on the Nomenclature of Biological Chemistry defines **Vitamin D** as a steroid or secosteroid. Actually we can consider it similar to an anabolic steroid hormone because research has shown that Vitamin D enhances the activity of the mTOR pathway and protein synthesis. In addition, it may boost testosterone levels. Recent research has identified VDRs (Vitamin

D receptors) in skeletal muscle, suggesting a direct relationship and effect between it and muscle. Related to this, Vitamin D deficiency has been linked to sarcopenia. So – while Vitamin D is not produced in the body as a consequence or response to exercise, it appears to play a role in muscle growth, which in my mind warranted its inclusion here.

Muscle Adaptations

The chief adaptation that comes to mind regarding training effects on muscle is growth. This phenomenon is linked to muscle damage incurred through overload training, and is a repair/rebuild /remodel response. In a previous section on muscle hypertrophy, we discussed the role of satellite cell activation, with the roles of the MRFs in myogenesis. These are all part of muscle adaptation to exercise.

Skeletal muscle cells (also called myocytes, muscle fibers or myofibers) are uniquely multinucleated cells in the body that assemble to form muscle tissue. Muscle tissue exhibits the ability to contract due to the presence of contractile proteins actin and myosin that exist as microfilaments within the myofibril bundles that make up each muscle fiber. Muscles function overall to provide body movement – to the extent that body motions are the result of the muscular system moving the body's skeletal structure (arms, legs, torso) or soft tissue structures (eyelids, mouth, and so forth). Fundamentally, muscles adapt to training by growing and increasing their contractile force or power output.

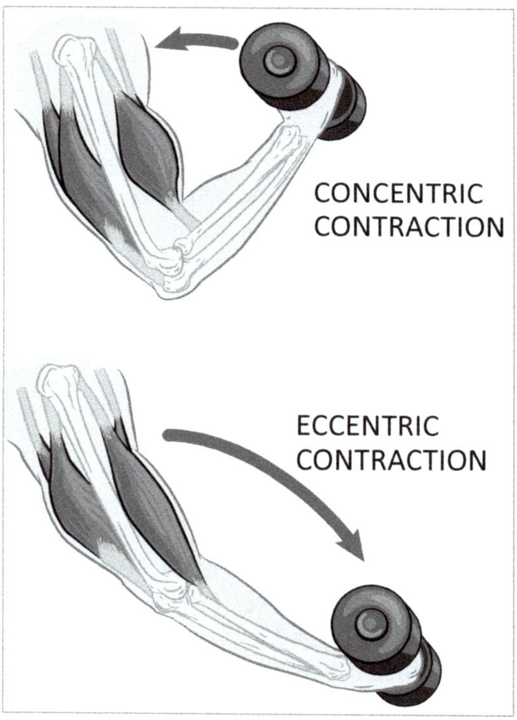

▲ **Figure 3.8** Concentric, eccentric, and isometric muscle contractions. In the concentric phase of an exercise, the weight is lifted as the muscle shortens. In the concentric phase of an exercise, the weight is lowered as the muscle lengthens. In the isometric phase of an exercise (not shown), the weight is held steady, and the muscle is often consciously and forcefully flexed, without moving up, down, or in any direction. Another way to envision this is to imagine yourself pushing against an immovable object, like a wall. Your muscles would be contracting isometrically. Concentric and eccentric contractions are said to be dynamic because there is movement, while isometric contractions are static.

Muscles have the unique ability to produce force when they shorten, lengthen, or hold steady. These three muscle movements or contractions are termed **concentric**, **eccentric**, and **isometric** contractions, respectively. They provide a diversity of ways to train to achieve the desirable adaptations of strength gain and also hypertrophy. It is all made possible by the fact that muscle tissue is responsive to the external resistance loads inherent in strength training, to which they adapt by becoming bigger and stronger. Let's not forget that anatomically, muscle connects to bone via tendons and ligaments. These soft tissues respond to training but in a manner different from muscle. They can strengthen, but only slowly. Also, if the resistance load is too heavy or the training too intense, they can actually become damaged and weaken. Bone itself responds to strength training by increasing in density, which is highly desirable in that the risk of spontaneous bone fractures (think osteoporosis) is reduced.

With resistance training, there are changes in skeletal muscle that specifically reflect fast twitch muscle fiber recruitment. At high enough intensities, these type II muscle fibers are recruited to a greater extent, with type I fibers less recruited. When we strength train, there is anatomical adaptation in the muscles that we are working, which leads to greater and at some point maximum muscle strength and size development by a certain age. Assuming we continue to train indefinitely, and support our training with proper nutrition and adequate rest and sleep, we will experience a period of muscle maintenance, before any final decline due to very advanced age. This is because it appears that if we live long enough, we will reach a point where gaining muscle mass and strength will plateau and then diminish in spite of our best efforts. This age will vary per individual, and is based upon a variety of factors (anabolic resistance to exercise and protein, sarcopenia, decreased mTOR action, decreased satellite cell activity, decreased anabolic hormones, etc.). As far as I can tell, no person in their 90s ever went to the grave looking ripped and huge like the Hulk! Possible exception: Jack Lalanne!

In a moment I will discuss **nervous system adaptations** and executive function, and there is one substance, BDNF that I will bring up at that time. But perhaps I should do so now, as it turns out that BDNF is relevant to muscle cells and not just brain cells. If you recall what we learned about satellite cells, one thing had to do with the manner in which they differentiate and express various myogenic factors. Well it turns out that **BDNF** (**brain derived neurotrophic factor**) exists not only in the brain but also in muscle. What is it doing there? Research has identified a role for **BDNF** (which is a form of cytokine called a myokine) in the regulation of satellite cell behavior and muscle regeneration. In addition, research has shown that BDNF produced by muscle remodels the neuromuscular synapses, the neuronal junctions between the motor neurons and muscle, and not only causes the Type II muscle fibers to develop, but endurance muscle fibers to decline. The type I fibers do not simply disappear; they are changed into type II fibers - proof of "muscle plasticity". Well that is certainly very, very interesting.

What about EPOC?

EPOC – excess post exercise oxygen consumption – can impact body weight over time, and be part of weight loss. It is related to the fact that muscles continue to be metabolically active for hours after a bout of exercise – especially a strength training workout. EPOC refers to the increased rate of oxygen uptake – hence energy expenditure - after working out. Some call it "the afterburn effect" because calories are being burned after the exercise session ends due to sustained metabolic elevation. Regarding weight loss, the thinking is that after strength training, the body requires time and energy to return itself to the pre-exercise state on multiple levels. For example, consider that the remodeling of muscle takes place in the time period between workouts, and this process of muscle building requires energy. Regarding replenishing ATP used up during the workout, the body will continue to use the aerobic (fat-burning) energy pathway to replace the ATP consumed during the workout for hours post-exercise. For these reasons, it is not an exaggeration to say that the more muscle mass a person has, the greater their ability to burn calories on a daily basis – during their sleep!

Nervous System Adaptations

Another type of adaptation occurs in the nervous system, in which neurons (nerve cells) and the muscle cells that they communicate with become "recruited" as assemblies called motor units. When this happens repeatedly in the presence of adequate nutrients, there can be muscle growth due to MPS – muscle protein synthesis. Increases in muscle tissue can be considered as hypertrophy (increase in muscle size) and hyperplasia (increase in cell number). Let's not forget that fundamentally, nerves and muscles are linked and work together to build strength. The motor neuron is physical evidence of the connection between the two cell types, which connect at **synapses**. The movements and force output of the different muscle fiber types all begin with engagement of the nervous system. Without the signal it provides to muscle, there can be no movement of muscle: no lifting, and no training. Exercise scientists refer to the sequence of events that start with a motor nerve impulse and result in muscle contraction as **excitation–contraction**.

I will present information on three areas related to adaptations and changes in the brain and CNS that can occur when we strength train properly. One is improvement of executive function, the second is development of the mind:muscle connection, and the third is the establishment of **mental well-being**. During exercise, the brain has the ability to change itself, a phenomenon known as **neuroplasticity**. This change includes the formation of new neural connections, and even new neurons - brain cells, a process called **neurogenesis**. I recall being told as a kid not to drink alcohol because we are born with a finite number of brain cells, and the alcohol will destroy them each time we drink, decreasing our ability to think and function. Well the toxicity of alcohol in undeniable, but the fact is that the brain can adapt, repair itself and create new brain cells. Not only brain function but also structure can be changed. And strength training can play a role in this.

Executive Function

A moment ago I mentioned BDNF in connection with muscle. But since this factor is called "brain derived', that means it also is present in brain tissue - where it does what? Answer: enhances neurogenesis and cognitive functioning – which brings us to the topic of executive function. What is meant by executive function? According to well-respected exercise researcher Len Kravitz, **executive function** refers to the brain processes that help us manage our day and plan activities. It includes the use of cognitive skills to coordinate behaviors such as planning new tasks, organizing information and making decisions.

Analogous to the manner in which exercise develops muscle, it also appears that exercise also stimulates neurological functioning. For example, brain activity data shows increased activation in memory-related areas of the brain immediately following physical activity. Research is increasingly demonstrating that many forms of exercise, including strength training, are cognitively-engaging activities. Physical activity has been shown to enhance cognition (the process of acquiring knowledge), memory, attention and academic performance.

The hippocampus region of the brain plays a crucial role in memory and learning. These two cognitive processes often see decline as we age. In addition, the ability to maintain one's balance, a function essential for well-being and quality of life, is hippocampus-associated. Once again, exercise can help maintain these critical functions by helping to maintain the function of this key brain area. Perhaps to no one's surprise, BDNF turns up again in the story; it has been found in high levels in the hippocampus and may play a vital role in learning and memory.

Mind: Muscle Connection

As we begin this next section on a very special phenomenon, a quick word about "mindfulness". **Mindfulness** is basically paying attention to what you are doing. It is concentrating and focusing, which when we train, is what we must do in order to lift properly with the correct form and execution. Which brings us to the topic of the mind: muscle connection. Just the fact that we can think about moving certain ways during training, and make our arms, legs, torso and so forth move in the manner in which any particular exercise demands, speaks to a **mind:muscle connection**. If I need to perform dumbbell curls as part of my workout, my conscious thought is to reach down and pick up a pair of dumbbells. In other words, our skeletal muscles are under the voluntary control of our minds. Which in theory means that we can exploit and develop this reality through our training to improve the results of our workouts.

We have learned a bit about motor units, so we already appreciate the importance and necessity of mind (the neuroanatomical part) and muscle working together properly during training in order to get the best results. In order to make our first year of strength training a pleasant experience rather than one of frustration, we of course want to see positive change by journey's end. And if the mind:muscle connection is present and working, we will. But if not – we will be going through the motions of working out without obtaining the best results. We can't just "swing the weights" and expect good things to happen. There has to be effort, true - but also overload, proper form, focus, consistency, and connection.

We should be able to realize right away that the mind:muscle skill that we can develop will work not only for isolation exercises when we want to focus on a single primary (target) muscle, but also when we desire to work multiple muscles at once. What we do NOT want to happen in isolation (single muscle) exercises is to let the secondary, supporting muscles "take over" due to sloppy technique and poor mind:muscle focus. This would minimize the effectiveness of the workout for the primary muscles, and we would not experience optimal growth in them. We have to focus our mind on the muscle we are mainly interested in working, and contract it with mind as well as with the muscle itself by thinking of it and feeling it contracting as we exercise it.

On the other hand, compound exercises are the foundation of strength training. These involve working multiple muscle groups in concert on purpose – which provides the best overall bang for the buck regarding the time spent in our workout and the whole body results. This is not only with respect to strength gains, but in terms of weight loss, boosting testosterone, fighting aging, and more. For this kind of exercise to be beneficial, we do NOT want to allow any one muscle group to dominate during the exercise. We mentally need to prepare to contract ALL the involved groups more or less equally in turn. Some are more important during the early contraction, some in the later contraction (toward the end of the concentric phase), and others in the eccentric phases. If we can train ourselves to think about these muscles and their sequence of action and contractions, and apply the mind:muscle approach (thinking, feeling, and doing) to our lifting behavior, we will be most likely to obtain the fullest benefit of any compound exercise.

▲ **Figure 3.9** The mind:muscle connection is important to understand and develop when strength training.

For example, consider the goblet squat that we perform in the Old Man Muscle workout. This is a compound exercise that works multiple muscles: the quads, hamstrings, glutes, and back. Holding and squeezing the dumbbell at the chest the way we do for this exercise (crush grip) also brings the arms, shoulders, and chest into the picture. So, in order to perform this exercise to benefit us the most as a compound exercise, we need to be sure to feel and activate all of these muscle groups as we perform the exercise in the sequence that they are most needed – which is a lot to think about in preparation! But it is essential to do so, otherwise we can fall into the trap of just going through the motion of the goblet squat and focus mostly on working the quads. Instead, feel and flex your back, feel and flex your arms, feel and flex your chest, glutes, quads – and all the rest – and this will be the great whole body exercise that it is meant to be, and you will feel like you have been run over by a truck afterward – a very good thing in this case!

And so it is also with another great whole-body compound exercise called the thruster. Is this one also a part of the Old Man Muscle workout? You bet it is! And I hasten to add that, as with the dumbbell version of the bench press that I recommend, we should use a grip on the dumbbells that is underhanded (supinated) rather than overhanded (pronated), which will significantly reduce the risk of overstressing the vulnerable shoulder joints, and ease any potential pain during the execution of these exercises. At the same time, the full benefit of the exercise to the target muscle groups will be achieved.

The bottom line is if we want to work out a particular muscle or group of muscles to maximum benefit, we need to make sure we develop a good mind-muscle connection skill. Otherwise, we'll never really be able to work out properly no matter what we do – the wrong muscles will take over and the ones we wanted to involve will be mere bystanders. The end result: the muscles we are trying to grow will not grow like we expected.

Lastly, it is generally true when we apply a good mind:muscle technique, that we can feel a muscle or muscle groups working due to a burning sensation in those muscles. So for example, in curling, although the wrists, forearms, triceps, shoulders, and back all can get involved to varying degrees, we want to minimize this potential, and place the emphasis on the biceps alone. If we succeed, they will start to burn as we complete our exercise. If this does not occur, then our mind:muscle is not developed for the biceps for this exercise and therefore it is not working – other muscles are stealing the workout. Of course this assumes that we have selected the proper weight and number of reps – just heavy enough to make eight or ten reps possible when it starts to burn. If not, what we would have to do is go back to the beginning and re-learn the exercise, the correct way. This would be unfortunate because we will have potentially already established millions of nerve:**muscle adaptations** and also **mental engrams** – all of which would need to be unlearned and replaced; this would take time. Not impossible, but time consuming and wasteful regarding our training schedule and progress. The bottom line: learn the exercises correctly from the start, and move ever forward – even if slowly, using a moderate resistance load for safety's sake!

Mental Well-being

The last benefit I will briefly mention has to do with the role of exercise in promoting positive mental well-being, by which I mean the decreased tendency toward anxiety and depressed mood. Strength training has specifically been associated with mood improvements in depression, loss of interest in activities, and feelings of worthlessness. Such effects are possibly as a result of increased blood flow to the brain when training, or the release of endorphins - norepinephrine and dopamine – that create a positive, upbeat mood. Anecdotally, I recognize in my own training a feeling of happiness during my workouts, not only when they are completed. There is an undeniable "high" that I experience, and I look forward to the next workout. Although exercise-induced endorphin release may be a way to support both the motivation and maintenance of regular exercise, it is believed that the intensity of the workouts can overrule positive motivation. By this I mean that when the workouts are habitually intense and demanding (think HIIT), the result can actually be discouragement and a lack of desire for the next workout due to negative feelings of pain, discomfort, and excessive challenge. Add to this the extra recovery demands, and it is easy to see why too much intensity is not advisable for older men from both motivation as well as maintenance perspectives. But let's not lose sight of the take-home message: it is clear from many studies that lifting weights is good for both the body and the mind. Strength training can help improve not only our physical but also our psychological health. And we can take advantage of the mind:muscle connection to get more out of our workouts.

ANABOLIC vs. CATABOLIC TRAINING

During strength training, there is stimulation of both the **catabolic** (breakdown) and the **anabolic** (synthesis) hormones of the endocrine system. And, while muscle is broken down during training, it is rebuilt – bigger and stronger, on recovery days. But as we age, the anabolic hormones that help build muscle are secreted in lesser and lesser amounts, while the catabolic ones remain. Examples of catabolic hormones are adrenaline, epinephrine and cortisol; anabolic examples are growth hormone (GH), IGF-1 (insulin growth factor one), and testosterone (T). Superimposed upon these endocrine realities is the anabolic resistance of older age to protein and to exercise, as well as sarcopenia. And then there is the issue of inflammation – which is another condition that acts against anabolic gains. It really does appear that the deck is stacked against us as older men, and that "anabolic insensitivity" or "anabolic blunting" looms as a real hurdle standing in our way. Such blunting means that muscle protein synthesis (MPS) and the master regulator for it (the protein kinase enzyme mTOR) become harder to activate the older we become. Remember, MPS is required for muscle building, and mTOR controls the anabolic and catabolic signaling of skeletal muscle that results in MPS. Therefore, anabolic training is our goal – we want there to be more building of muscle than tearing down, providing the stimulus for MPS and for mTOR activation. But how can we achieve this?

The classical interventions to counteract the rise of catabolic hormones, anabolic resistance, sarcopenia, and the detrimental effects of inflammation have centered upon resistance training with weights, along with increasing the amount of dietary protein. I want to focus now on the manner in which we strength train, suggesting that it is possible to create a workout that is generally more anabolic, and that we should avoid exercising in a manner that is overly catabolic. In other words, we need to exercise anabolically with weights, and take in adequate dietary protein, in order to stimulate MPS and mTOR. As a warning, if we over-train, we get into a catabolic state with our working out – the result of which is that we fail to build muscle and actually lose muscle mass over time. Similarly, endurance training is typically significantly less anabolic than is strength (resistance) training with weights. We need to spend less time on the treadmill and more time focusing on weight training.

Before I go any further, it is generally accepted that anabolic exercises (strength training, power workouts) and catabolic exercises (aerobics, endurance workouts) complement each other and can be part of a wise training regimen, especially in the younger age groups that can engage in HIIT (high intensity interval training). Performing both anabolic and catabolic exercises provides variety, and helps burn calories while building muscle and losing fat. However, there is risk involved for older adults contemplating HIIT because of the stress of the exercise pace, the intensity level, risk of injury, increased recovery time needed, and so forth. The intensity of HIIT for an older person might be possible but for relatively few workouts, in my opinion. I see no advantage to HIIT training for seniors because the result will almost certainly be overtraining syndrome (OTS) and injury if it is attempted long-term. Instead, it is more prudent - and anabolically beneficial – to focus on strength training our large muscle groups and the joints that they are associated with – rather than intense cardio and aerobic training.

In order to train anabolically as older men, we need to emphasize multi-joint, large muscle group lifts such as squats, deadlifts, dumbbell swings and thrusters, and supplement these with well-chosen upper and lower body exercises. Although we can perform aerobic exercises like walking, jogging, cycling and swimming on days we are not lifting, our emphasis as we age should be on strength training. This not only will provide the anabolic boost, but also help us better perform our ADLs (activities of daily living) into old age. I also want to mention that there is a difference between slowly riding a bicycle at a relaxed pace on a sunny day around the neighborhood compared to intensely cycling for 20 minutes on a stationary bike. The former is a very light exercise activity and will not be particularly anabolic or catabolic, while the latter can be of such an intensity that it is primarily catabolic in outcome. The telltale clue is how much the heart rate is raised, how your breathing is affected, and pain and fatigue factors. There can also of course be negative effects on the muscles and joints if we exercise in an unwise manner.

Although there are many ways to strength train, we need to do more than body weight exercises and elastic band exercises, although these can be great at the beginning - especially for the elderly and frail. We need to overload our muscles through the use of free weights in our strength training. This is why I incorporate such exercises into the Old Man Muscle workout. The Old Man Muscle workout is an anabolic workout. The exercises it features are likely to stimulate anabolic growth hormones significantly more than do endurance type activities. This results in greater increases in protein synthesis and muscle mass (hypertrophy) than does endurance training. I will make the point in the next chapter that the Old Man Muscle workout can work for just about anyone (my wife, for example!), but I am not claiming that it is somehow magical or special compared to other training workouts. It is probably better than some, equal to many, and not as effective as some. But the bottom line is that if you are an older man looking to get in shape after decades of neglect, strength training is the way to go, and I've put something together that worked for me that I am excited to share.

The actual proof that you are successfully training anabolically is the visual evidence of muscle growth. Look in the mirror at the start of your training, and compare what you see in 3, 6, 9, and 12 months. If you train consistently, follow a good diet with enough protein, and get enough rest between workouts and quality sleep each night, you should be smiling as the months go by. Visual muscle growth is also evidence that your anabolic hormones are being activated by your strength training, and that your dietary protein is adequate to meet the demands of training and is successfully overcoming any barriers (e.g. anabolic resistance and so forth). You may have worried about your T levels or GH levels – but if you are putting on muscle as you train, there is clearly nothing to be concerned about! You could not be achieving results if these anabolic hormones were in the tank.

Training, DOMS, Overtraining

Muscle fatigue is a reality of training, and in fact is an indicator of both good things and not-so good things. By this I mean, if we are not routinely training our muscles to fatigue, then we will not gain the increase in muscle size (hypertrophy) and strength that are primary goals of our workout. If we do not push the muscles to the point where they will start to release a host of metabolic products, we will fail to trigger MPS (muscle protein synthesis), and fail to send out signals that communicate with other tissue including adipose (fat) and brain, exerting very positive effects in those organs (fat burning and neurogenesis – making new brain cells, respectively). Both pro- and anti-inflammatory signals are also generated when we train, rest, and eat properly. In fact, we can establish a situation in the body that is predominantly anti-inflammatory, which is highly desirable. So these are the good things related to fatigue.

But when is fatigue a negative? Simply put, when fatigue is allowed to occur that is not followed by adequate rest between workout days, or we work too intensely and become extra-fatigued, this is not good. It opens the door to (1) increased risk of injury, (2) the inability to work to our fullest potential, and (3) is a sign of overtraining – something we never want to do.

Although mentioned previously, I want to again speak about **DOMS**. This refers to muscle tissue damage and probably also inflammation – resulting in post-workout pain and muscle weakness - when a person unaccustomed to resistance exercise starts training. DOMS, delayed onset muscle soreness, is a normal response to overload training. It is a sure sign that you are placing enough stress on the muscles to initiate a growth stimulus within 24-36 hours. DOMS is not pain during exercise. DOMS requires time to establish after working out (often several hours). I have noticed in my own training that for up to a day after each workout, I will often feel that my muscles are weakened (fatigued) and would not want to be called upon to do heavy lifting, but there is not always associated pain. So for some, DOMS may emphasize temporary muscle fatigue rather than pain.

What causes DOMS? It has been asserted that **myoglobinemia** (uptake of muscle myoglobin into the blood), and a condition known as rhabdomyolysis – or rhabdo – are to blame. This happens when certain stress-related metabolites and by-products of tissue damage besides myoglobin are generated (recall metabolic overload training that produces lactate and H+ ions) during intense activity. It has been reported that there is no specific effective treatment for DOMS pain, although there are many claims. Mild heat may help some DOMS sufferers, as well as analgesics. Fortunately DOMS typically resolves on its own in a couple of days, give or take. It is simply part of the recovery and rebuilding process of strength training.

On the other hand, if you experience pain during training, it could be serious in that soft tissue (a tendon, ligament, or muscle) have been strained or possibly even torn. In this case, you should stop immediately and follow the rules of **RICE**: rest or at least restricted activity, ice applied every hour for 10-20 minutes to the injured area to reduce swelling and inflammation, compression (apply a compression bandage or wrap around the area), and elevation of the area above the heart.

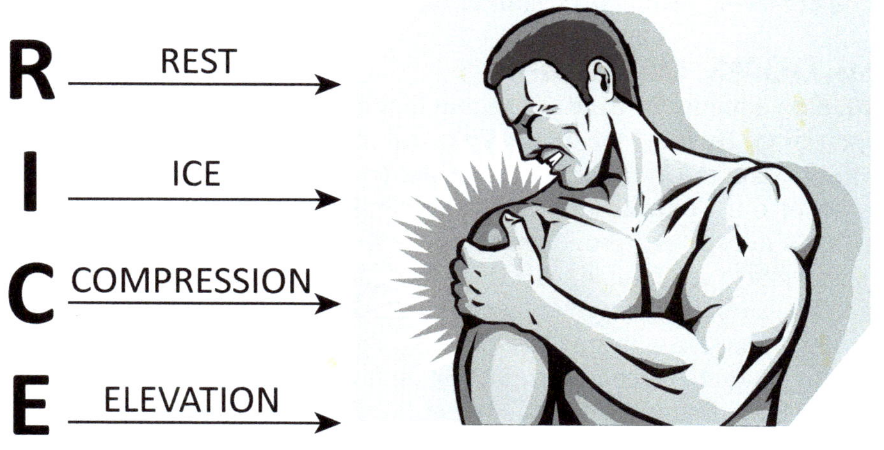

▲ **Figure 3.10** Treatment for soft tissue injury involves RICE (rest, ice, compression, elevation). In the event musculoskeletal injuries, medical management is often the course to follow. Typical treatments may include cortisone injections, immobilization of affected part, oral anti-inflammatory agents (NSAIDs), physical therapy, and sometimes surgery.

In the workout routine, we work certain muscle groups initially, and allow them to rest as we deconstruct the remainder of the workout. What I mean to say is, we try not to line up all our arm exercises at the start, because this will cause mega arm fatigue, and we won't have enough energy left to compete the rest of the workout. So we mix up the routine so that muscles are called upon to literally do the heavy lifting early, then can rest and play a background role in a few exercises before being called upon again for primary use. We want to get the most challenging lifts done first while we are fresh. The funny thing is, as remarkable as muscles are, even with professional lifters, muscles become fatigued in a matter of moments depending on what exercises and poundage are thrown their way – and that's just the way it is.

Lifting and training the way we will be doing is not an activity that the muscles can crank out for 60 minutes. Fatigue and DOMS are a sure thing, and as we have stated, a good thing. They are not the same as pain due to injury. So we must make the most of the time that we are in the gym, or training at home, as the case may be. Once the workout is over for the day, there is no "going back" to redo it. The muscles will already be optimally fatigued and if we try to do more that is when overtraining-related injuries can happen. We must immediately allow for rest and recovery until the next workout, typical spacing one day of rest (two is also OK) between workouts.

As we get ready to conclude this chapter, the issue of **overtraining** is important to mention. We all know the trap that we can fall into when we think, "if a little is good, more must be better". It is true that when we exercise, we create adaptations that are very beneficial to health, such that "exercise is medicine" rings as a true statement. But when it comes to training volume, there is a dose-response relationship, which means there is a maximum point after which things go downhill. This is what overtraining is all about: doing too much for too long without allowing for ample time off for rest and recovery, and this includes taking in sufficient fuel in the form of protein and carbs. If you are training yourself, it is on you not to get too enthusiastic and overdo it. If you have a trainer, it is on them to control your training volume properly.

Realize too that research is only beginning to answer questions related to strength training and building muscle in older men, when it comes to the key training variables that can be controlled. These include identifying the optimal resistance load as % 1-RM, the number of reps, the number of sets, the pace of the lifting, the rest time between sets, the total training volume, and the frequency of training per week. Theoretically, we can add to this such things as mixing and matching variations of mechanical tension, metabolic stress, and muscle damage effects in the training approach, as well as training with a split routine or a whole body workout. It is as yet undetermined what the ideal situation is in order to maximize strength and hypertrophy results, certainly for older men but also for any demographic group, trained or untrained. For the reader who is keenly interested in such things, I would suggest reading published studies by internationally renowned researcher Dr. Brad Schoenfeld, who is investigating the cutting edge of muscle growth and training parameters (see if you can find out what **DUP**, or daily undulating periodization is all about!).

▲ **Figure 3.11** Some often-observed symptoms associated with overtraining, or OTS (overtraining syndrome).

NUTRITIONAL SUPPORT FOR TRAINING

Pre-, Peri-, and Post-workout Nutrition

TIMING	SUGGESTED FOODS AND BEVERAGES
PRE-WORKOUT	Water; slow carbs; protein foods that are low in fat (low-fat cottage cheese with some grapes or berries; oatmeal topped with nonfat milk; high protein cereal with skim milk; tuna or chicken breast on whole grain bread; one whole fruit - apple, banana, peach - plus a protein beverage, etc.); no sugary or salty snack foods
PERI-WORKOUT	Focus on water intake
POST-WORKOUT	Water; carbs containing fructose replenish liver glycogen, carbs containing glucose replenish muscle glycogen; fruit juices (apple, cranberry, grape are suggested); EAAs from high quality protein foods or protein beverages will contribute significantly to MPS

▲ **Table 3.8** Text to come

Before you start training, it is wise to have a plan in place regarding the nutritional support your workouts will require. The importance of what we eat on a daily and weekly basis, and beyond – for life - is of course one of the main messages of this book. In Chapter 5, the role of nutrition as a lifestyle behavior is examined. But there's something to be said for the importance of proper nutrition to complement each individual strength training workout that we engage in. In fact, a great question to ask involves the timing of food intake with respect to training on workout days. This topic, nutrient timing, is typically addressed in terms of pre-workout (foods eaten before training), peri-workout (foods eaten during training), and post-workout (foods eaten after the training session). As we consider each in turn, let's realize that although we are saying "foods" here we surely don't want to forget water!

Pre-workout nutrition refers to what we should intentionally intake approximately 2-3 hours prior to strength training. Such pre-workout nutrition must be handled carefully, as it will directly impact our ability to perform during our workout. Fast ATP energy can be supplied to the brain and working muscles (remember, brain and muscle coordinate during training) through intake of carbohydrate. On the other hand, protein is needed for its anabolic and anti-catabolic effects, while water will provide hydration to body cells. Water prior to training is absolutely essential; it is believed that some caffeine is beneficial at this stage as well.

Peri-workout nutrition refers to what we consume during our workouts. The main thing here is water. Depending upon a variety of factors, we will lose a certain amount of water through sweat and respiration during strength training. So, we need to intake water throughout our time of exercise in order to rehydrate. The other nutrients that could potentially assist in the peri-nutrition timeframe are carbohydrate and protein. However, if the pre-nutrition meal has been consumed and already provided these, intaking more at this point is mostly not necessary. In fact, peri-nutritional strategies are more for folks engaged in lengthy, marathon exercise as well as intense, extended physical activity – which is typically beyond the boundaries of how we should strength train as older men.

Post-workout nutrition refers to what we should intentionally ingest within 0-2 hours after each workout is completed. This requires water again – for rehydration purposes, along with some carbohydrate and a heaping helping of protein. Regarding protein, it used to be believed that there was a very narrow "anabolic window" that closed in less than one hour after completing a workout. In this scenario, failure to rush to consume for example a protein shake within minutes was believed to hinder post-exercise MPS. However, recent research has indicated that the timing of nutrient intake post-workout can be up to at least two hours following training. For peace of mind, sure – see to it as soon after the workout as possible. But don't feel that you've missed the boat if you are delayed for an hour or two, as has happened to me on several occasions. Research currently suggests consuming between 20-50 grams of protein post-workout, depending upon our body weight. The protein choice should be of high quality, providing all of the EAAs (essential amino acids). Manufacturers formulate convenient-to-consume beverages, or powders that can easily be added to water or milk for such purposes.

RECAP AND WHAT'S AHEAD

The ambitious goal of this chapter was to catch us up on a lot of detail regarding what we should know before we start training. It spanned everything from goal setting to strength training basics, muscle hypertrophy to training intensity, and training adaptations to DOMS. A lot of the information became technical and quite detailed! This may mean the need to read over the material multiple times to make sense out of it. But the pay-off is straight ahead. Armed with this background, we progress to our next chapter to learn how to perform the specific **exercises of the Old Man Muscle workout**, so as to begin the process of reshaping and strengthening our aging bodies.

Anton MM, et al. 2006. Resistance training increases basal limb blood flow and vascular conductance in aging humans. Journal of Applied Physiology 101:1351–55.

Aragon AA, Schoenfeld BJ. 2013. Nutrient timing revisited: is there a post-exercise anabolic window? Journal of the International Society of Sports Nutrition 10:5.

Arsic N, et al. 2001. Vascular endothelial growth factor stimulates skeletal muscle regeneration in vivo. Mol Ther.10: 844–854.

Aristizabal JC, et al. 2015. Effect of resistance training on resting metabolic rate and its estimation by dual-energy X-ray absorptiometry metabolic map. European Journal of Clinical Nutrition 69(7):831–36.

Asfour HA, Allouh MZ, Said RS. 2018. Myogenic regulatory factors: The orchestrators of myogenesis after 30 years of discovery. Exp Biol Med (Maywood) 243(2):118–128.

Baechle T and Earle R. 2008. Essentials of Strength and Conditioning, 3rd edition. Champaign, Ill. Human Kinetics.

Baty JJ, et al. 2007. The effect of a carbohydrate and protein supplement on resistance exercise performance, hormonal response, and muscle damage. J Strength Cond Res. 21(2):321-9.

Beck DT, Casey DP, Martin JS, et al. 2013. Exercise training improves endothelial function in young pre-hypertensives. Experimental Biology and Medicine, 238(4):433-441.

Beck DT, et al. 2013. Exercise training improves endothelial function in young pre-hypertensives. Experimental Biology and Medicine 238(4):433–41.

Berardi JM, et al. 2006. Postexercise muscle glycogen recovery enhanced with a carbohydrate-protein supplement. Med Sci Sports Exerc.38(6):1106-13.

Bersheim E, Bahr R. 2003. Effect of exercise intensity, duration and mode on post-exercise oxygen consumption. Sports Medicine 33(14):1037-1060.

Bubbico A, Kravitz L. 2011. Muscle hypertrophy: New insights and training recommendations. IDEA Fitness Journal 2326.

Bweir S, Al-Jarrah M, Almalty A-M, et al. 2009. Resistance exercise training lowers HbA1c more than aerobic training in adults with type 2 diabetes. Diabetology & Metabolic Syndrome. Accessed at: https://www.ncbi.nlm.nih.gov/pubmed/20003276

Bweir S, et al. 2009. Resistance exercise training lowers HbA1c more than aerobic training in adults with type 2 diabetes. Diabetology & Metabolic Syndrome 1(27).

Centers for Disease Control and Prevention. National Diabetes Statistics Report, 2017. Atlanta, GA.

Centers for Disease Control and Prevention. 2019. High Cholesterol Facts. Accessed at: https://www.cdc.gov/cholesterol/facts.htm

Cornelissen VA, Smart NA. 2013. Exercise training for blood pressure: a systematic review and meta-analysis. Journal of the American Heart Association,

de Sousa EC, Abrahin O, Ferreira A.L, et al. 2017. Resistance training alone reduces systolic and diastolic blood pressure in prehypertensive and hypertensive individuals: meta-analysis. Hypertension Research ; doi:10.1038/hr.2017.69

Cotman CW and Engesser-Cesar C. 2002. Exercise enhances and protects brain function. Exercise and Sport Sciences Reviews 30(2):75–79.

Crewther C, et al. 2006. Possible stimuli for strength and power adaptation: Acute hormonal responses. Sports Medicine 36, 3, 215238.

Delezie J, et al. 2019. BDNF is a mediator of glycolytic fiber-type specification in mouse skeletal muscle. Proceedings of the National Academy of Sciences.

Accessed at https://www.sciencedaily.com/releases/2019/07/190725112607.htm

Demer LL, Hsu JJ, Tintut Y. 2018. Steroid Hormone Vitamin D: Implications for Cardiovascular Disease. Circulation Research. 122(11). Accessed at: https://www.ahajournals.org/doi/full/10.1161/CIRCRESAHA.118.311585

Erem S, Atfi A, Razzaque MS. 2019. Anabolic effects of vitamin D and magnesium in aging bone. Journal of Sterioid biochemistry and Molecular Biology. Accessed at: https://www.sciencedirect.com/science/article/abs/pii/S0960076018307362

Esmarck B, et al. 2001. Timing of postexercise protein intake is important for muscle hypertrophy with resistance training in elderly humans. J Physiol. 15;535(Pt 1): 301-11.

Etnier, J, et al. 2016. The effects of acute exercise on memory and brain-derived neurotrophic factor (BDNF). Journal of Sport and Exercise Psychology 38(4), 331–340.

Fahlman MM, Boardley D, Lambert CP, et al. 2002. Effects of endurance training and resistance training on plasma lipoprotein profiles in elderly women. Journal of Gerontology 57(2):54-60.

Fisher J, Steele J. Smith D. 2013. Evidence-based resistance training recommendations for muscular hypertrophy. Medicina Sportiva, 17, 4, 217235.

Garcia D, Shaw RJ. 2017. AMPK: Mechanisms of Cellular Energy Sensing and Restoration of Metabolic Balance. Molecular Cell Review 66:789-800.

Goh Q, et al. 2019. Myonuclear accretion is a determinant of exercise-induced remodeling in skeletal muscle. Elife. Accessed at: https://www.ncbi.nlm.nih.gov/pmc/articles/PMC6497442/

Goodson A. 2018. Industry Presented Blog: What to Eat Before and After a Workout. Accessed at: http://www.acsm.org/blog-detail/acsm-certified-blog/2018/06/25/what-to-eat-before-and-after-a-workout

Greer BK et al. 2015. EPOC comparison between isocaloric bouts of steady-state aerobic, intermittent aerobic, and resistance training. Research Quarterly for Exercise and Sport 82(2):190–95.

Gunton JE, Girgis CM. 2018. Vitamin D and muscle. Bone Rep. 8:163–167.

Hales CM, Carroll MD, Fryar CD, et al. 2017. Prevalence of obesity among adults and youth: United States, 2015-2016. NCHS Data Brief, 288, 1-8.

Holten MK, Zacho M, Gaster M, et al. 2004. Strength training increases insulin-mediated glucose uptake, GLUT-4 content, and insulin signaling in skeletal muscle in patients with type 2 diabetes. Diabetes 53: 294-305.

Kadi F. 2008. Cellular and molecular mechanisms responsible for the action of testosterone on human skeletal muscle. A basis for illegal performance enhancement. Br J Pharmacol. 154(3):522–528.

Kerksick CM, Arent S, Schoenfeld BJ, et al. 2017. International Society of Sports Nutrition position stand: nutrient timing. J Int Soc Sports Nutr 14:33.

Kravitz L. 2012. Resistance training benefits mind and body. IDEA Fitness Journal.

Kreider RB. 2019. Essentials of exercise & sport nutrition: science to practice. Lulu Publishing Services.

LaForgia J, Withers R, Gore C. 2006. Effects of exercise intensity and duration on the excess post-exercise oxygen consumption. Journal of Sport Sciences 24(12):1247-1264.

Levenhagen DK, Gresham JD, Carlson MG, Maron DJ, Borel MJ, Flakoll PJ. 2001. Postexercise nutrient intake timing in humans is critical to recovery of leg glucose and protein homeostasis. American Journal of Physiology Endocrinology and Metabolism. 2001;280:E982–E993.

Liu Y, et al. 2018. Associations of Resistance Exercise with Cardiovascular Disease Morbidity and Mortality. Medicine & Science in Sports & Exercise. Accessed at: https://www.sciencedaily.com/releases/2018/11/181113115430.htm

Mann S, Beedie C, Jimenez A. 2014. Differential effects of aerobic exercise, resistance training and combined exercise modalities on cholesterol and lipid profile: Review, synthesis and recommendations. Sports Medicine 44(2):211–21.

Mayer F, Scharhag-Rosenberger F, Carlsohn A, Cassel M, Müller S, Scharhag J. 2011. The intensity and effects of strength training in the elderly. Dtsch Arztebl Int. 108(21): 359–364.

Mohamad NI, Cronin JB, Nosaka, KK. 2012. Difference in kinematics and kinetics between high- and low-velocity resistance loading equated by volume: Implications for hypertrophy training. Journal of Strength and Conditioning Research 26(1):269275.

Mozaffarian D, Benjamin EJ, Go AS, et al. 2015. Heart disease and stroke statistics-2015 update: a report from the American Heart Association. Circulation, 131, e29-e322.

Munoz-Martinez FA, Rubio-Arias JA, Ramos-Campo DJ, et al. 2017. Effectiveness of resistance circuit-based training for maximum oxygen uptake and upper-body one-repetition maximum improvements: a systematic review and meta-analysis. Sports Medicine 47(12):2553-2568.

Narayanasetti PTN and Thomas PT. 2017. Exercise and Neural Plasticity– A Review Study. J Neurol Neurosci. 8(5):216.

Northey JM, et al. 2018. Exercise interventions for cognitive function in adults older than 50: a systematic review with meta-analysis. British Journal of Sports Medicine 52(3):154-160.

Olsen S, et al. 2006. Creatine supplementation augments the increase in satellite cell and myonuclei number in human skeletal muscle induced by strength training. Journal of Physiology 573(Pt 2):525-534.

Olson TP, Dengel DR, Leon AS, et al. 2006. Moderate resistance training and vascular health in overweight women. Medicine and Science in Sports and Exercise, 38(9), 1558-1564.

Ormsbee MJ, Bach CW, Baur DA. 2014. Pre-Exercise Nutrition: The Role of Macronutrients, Modified Starches and Supplements on Metabolism and Endurance Performance. Nutrients 5(5):1782-1808.

Pilz S, et al. 2011. Effect of vitamin D supplementation on testosterone levels in men. Hormone and Metabolic Research 43(3):223-5.

Piras A, Persiani M, Damiani N, et al. 2015. Peripheral heart action training as a valid substitute to high intensity interval training to improve resting cardiovascular changes and autonomic adaptation. European Journal of Applied Physiology, 115, 763-773.

Robinsin J. 2016. What You Need to Know About Nutrient Timing. Accessed at: https://www.acefitness.org/education-and-resources/professional/expert-articles/5815/what-you-need-to-know-about-nutrient-timing

Salles J, et al. 2013. 1,25(OH)2-vitamin D3 enhances the stimulating effect of leucine and insulin on protein synthesis rate through Akt/PKB and mTOR mediated pathways in murine C2C12 skeletal myotubes .Mol Nutr Food Res. 57(12):2137-46Sambasivan R, et al. 2011. Pax7-expressing satellite cells are indispensable for adult skeletal muscle regeneration. Development 138: 3647–3656.

Schoenfeld BJ, et al. Effects of Varied versus constant loading zones on muscular adaptations in trained men. 2016. Int J Sports Med. 37(6):442-7.

Schoenfeld B. 2010. Pre-workout nutrition. Accessed at: https://www.lookgreatnaked.com/blog/pre-workout-nutrition/

Schoenfeld B. 2010. Post-workout nutrition. Accessed at: http://www.lookgreatnaked.com/blog/post-workout-nutrition/

Schoenfeld BJ, Aragon AA, Wilborn C, Urbina SL, Hayward SE, Krieger J. 2017. Pre-versus post-exercise protein intake has similar effects on muscular adaptations. PeerJ. Accessed at: https://www.ncbi.nlm.nih.gov/pmc/articles/PMC5214805/

Schoenfeld BJ, Ogborn D, Krieger JW. 2017. Dose-response relationship between weekly resistance training volume and increases in muscle mass: A systematic review and meta-analysis. Journal of Sports Sciences 35(11):1073–82.

Schoenfeld BJ. 2010. The mechanisms of muscle hypertrophy and their application to resistance training. Journal of Strength and

Conditioning Research, 24(10):2857–72.

Schoenfeld, BJ. 2013. Potential mechanisms for a role of metabolic stress in hypertrophic adaptations to resistance training. Sports Medicine 43, 179194.

Sibley BA, Etnier JL. 2003. The relationship between physical activity and cognition in children: A metaanalysis. Pediatric Exercise Science 15:243–256.

Spangenburg E. 2009. Changes in muscle mass with mechanical load: Possible cellular mechanisms. Applied Physiology, Nutrition and Metabolism 34, 328335.

Thompson DM. 2018. The Role of AMPK in the Regulation of Skeletal Muscle Size, Hypertrophy, and Regeneration. International J of Molecular Sciences 19:3125.

Timpani CA, Hayes A, Rybalka E. 2017. Therapeutic strategies to address neuronal nitric oxide synthase deficiency and the loss of nitric oxide bioavailability in Duchenne Muscular Dystrophy. Orphanet J Rare Dis. 12(1):100.

Tipton KD, et al. 2001. Timing of amino acid-carbohydrate ingestion alters anabolic response of muscle to resistance exercise. Am J Physiol Endocrinol Metab. 281(2):E197-206.

Tomporowski PD, Davis CL, Miller PH, Naglieri JA. 2008. Exercise and children's intelligence, cognition, and academic achievement. Educational Psychology Review 20:111–1

Verkhoshansky Y, Siff M. 2009. Supertraining, 6th edition. Rome, Italy: Verkhoshansky.

Vingren J, et al. 2010. Testosterone physiology in resistance exercise and training. Sports Medicine 40, 12, 10371053.

Weaver C. Rethinking recovery nutrition: what you eat before your workout affects your post-workout. Accessed at: https://breakingmuscle.com/healthy-eating/rethinking-recovery-nutrition-what-you-eat-before-your-workout-affects-you-post-worko

Wernbom M, Augustsson J, Thomee R. 2007. The influence of frequency, intensity, volume and mode of strength training on whole muscle cross-sectional area in humans. Sports Medicine 37, 3, 225264.

Westcott WL. 2015. Build muscle, improve health: Benefits associated with resistance exercise. ACSM's Health & Fitness Journal 19(4):22–27.

Willoughby DS, Stout JR, Wilborn CD. 2007 Effects of resistance training and protein plus amino acid supplementation on muscle anabolism, mass, and strength. Amino Acids 32(4):467-77.

Won J. 2019. Semantic memory activation after acute exercise in healthy older adults. Journal of the International Neurophysiological Society 25(6):557-568.

Yin H, Price F, Rudnicki MA. 2013. Satellite cells and the muscle stem cell niche. Physiol Rev. 93(1):23–67.

Zanou N, Gailly P. 2013. Skeletal muscle hypertrophy and regeneration: interplay between the myogenic regulatory factors (MRFs) and insulin-like growth factors (IGFs) pathways. Cell Mol Life Sci. 70(21):4117-30.

Zatsiorsky V, Kraemer W. 2006. Science and Practice of Strength Training, 2nd edition. Champaign, Ill.: Human Kinetics.

Zhang P, et al. 2015. mTOR is necessary for proper satellite cell activity and skeletal muscle regeneration. Biochem Biophys Res Commun. 463(1-2):102-8.

THE OLD MAN MUSCLE WORKOUT

General Principles

My friends, we need to take care of ourselves in our later years in ways that we didn't have to consider when we were young. Not only does our physical appearance and body change, but our priorities change, our metabolism changes, we have possible joint issues, teeth, heart, blood sugar, back problems, erectile dysfunction – quite a laundry list of woes. And, due to a variety of factors most often linked to poor lifestyle habits (smoking, drinking, excessive food and sugar/sweets intake, lack of physical activity) some of us truly are "walking heart attacks waiting to happen". The other obvious reason to take better care of ourselves is that we are all closer to our "expiration date" – and while I am not saying that our goal should be to "cheat the inevitable", we can take steps to extend our time on planet earth, being in the best possible functional health through sound lifestyle choices.

Well, this kind of disaster need not continue, nor ever be this way. That is because we have the ability to control much of what are referred to as "lifestyle chronic diseases" through a proper diet, adequate **sleep** and stress relief, plus strength training or other age-appropriate forms of physical activity. We don't need to resort to surgery or be on five different medications to manage what we often wrongly assume are conditions simply due to old age. Often such conditions are mostly due to poor lifestyle habits, and these can and should be corrected.

In this chapter, I will share with you the specifics of the Old Man Muscle exercise routine, or workout. But first I need to discuss a few things regarding strength training exercises. There are numerous exercises that can be incorporated into any given strength training program. Over the years, certain exercises have emerged as being the more conventional and traditional – for a variety of reasons. This does not necessarily mean that they are the absolute best in terms of building strength and muscle while avoiding injury, or that they are the most mechanically efficient and effective. There is the potential for any exercise, when wrongly performed, to fail in these respects. Although it is true that performing some exercise on a regular basis is better than none at all, it makes sense to select and perform exercises intelligently.

By using age-appropriate resistance loads delivered through conventional exercises employing a reasonable number of sets and reps at the start, the Old Man Muscle workout attempts to minimize injury to a vulnerable population (older men) while at the same time providing a basis for improving muscle strength and growth. Again, I am sharing the workout that I devised and used my first year of training that provided me with results – so I know from my experience that it works. This is not to say that other workouts are wrong or bad or won't work. I am simply explaining the rationale for my sharing this specific workout with the reader as presented within the pages of this book.

The Old Man Muscle workout has been designed to be safe, effective, and able to accomplish the desired fitness and personal goals of participants. This occurs in conjunction with the application of fundamental principles of biomechanics and kinesiology. Specifically, for our purposes, consideration of age-related biomechanical issues, such as compromised joints and muscle imbalances, is critical. While it is usually beyond the scope of the average person to be knowledgeable in these areas, that is where a personal trainer can provide assistance.

If an individual embarks on a strength training program with such issues that manifest as postural deviations and pain during certain body movements, then modified strength training exercises different from what are presented in this book may be indicated, as well as specific stretching exercises. Again, it would be advisable to seek advice from a knowledgeable personal trainer. For instance, rather than using free weights, it may make better sense to work out with cable and pulley strength training equipment, or even elastic bands - often in the seated position, as these are easy on the joints. Furthermore, regular stretching of the spine, shoulders, hips, glutes, and quads is desirable in order to overcome the stiffness associated with limited flexibility, limited mobility, and diminished performance in the gym as well as when performing daily tasks.

We definitely want to do something to counter the toll that aging and sedentary living have taken on us. Fortunately, there is a large body of evidence from reliable research sources regarding the many benefits of strength training. Not strength training merely to shape new muscles – but to achieve functional strength – so that we can accomplish the tasks of daily living more easily and with greater confidence, as we grow older. And this is where the Old Man Muscle workout finally comes in. There are specific exercises that we will perform that will help us get up out of a chair, climb stairs, reach for items on a shelf, go ten rounds with the neighborhood bully – just kidding about that one – but seriously, so much more. In fact, now comes the fun part – and an interesting one – trying to adequately serve as your fitness instructor through the use of a book and printed words and photographs, rather than in person. However, I think this can work, and I'm ready with the information if you are.

Before describing the actual workout itself, I wish to make a few points.

First of all, I am providing a sound plan that should enable participants to succeed in strength training when followed. I want to emphasize that in my entire life, I never participated in any strength training program, nor did I ever lift weights for more than two weeks on my own – I quickly grew tired of it and hated it. Plus, I was never a competitive athlete. I played basketball recreationally from roughly the age of 17 to 21 – and that was it. I started the sedentary lifestyle thereafter, with married life and a career, such that I essentially was a couch potato in my 30s, 40s, 50, and early 60s. Therefore I had no "muscle memory" for strength training when I turned to it in 2017, at age 63. I conceived of and started my Old Man Muscle plan in May of 2017, an old, out-of-shape man - and as I write this, I have completed two years of training. I am so happy with the results: I have lost weight, gained muscle mass, feel and look stronger (and possibly even younger, at least from the neck down) and I have lowered my blood pressure to that of a healthy younger man (as long as I'm not sitting in the dental patient chair!) In fact, I was just in the hospital visiting with a patient, and asked the nurse if she could take my blood pressure. It was 111/70 – not bad for an old geezer.

Following the Old Man Muscle workout is in contrast to joining a gym and waltzing in and out on certain days when you feel like it, lifting a few random weights here and there or using a few machines, with no structured plan of progressive resistance training to build muscle and strength efficiently, effectively, and according to schedule. It really pays to have a plan and to stick to it in order to achieve results. I am providing you the plan here that has worked for me. Even though results may vary from person to person due to a variety of factors, if you apply it consistently, it should work well for you too.

Second, this is an age-appropriate strength training and muscle-building workout. I have selected exercises and grouped them in a sequence that makes sense on a variety of levels, including being suitable to achieve our goals of building muscle, developing strength, and doing so using a routine that is safe for men aged in their 50s, 60s and older. The sequence is such that the most challenging lifts come at the beginning when we are at our freshest, and the perhaps more fun ones come toward the end when we are becoming fatigued but will be motivated to finish with a fun lift rather than an intimidating lift. If a particular individual cannot perform all of the stretches and exercises, that is possible and understandable – not everyone will be able to for one reason or another at the outset. However, research continually has shown that doing some exercise is better than doing none! You will still receive benefits from this plan, even if you cannot perform say, deadlifts or squats, for example. Importantly, it is essential to realize that we cannot begin this workout without first warming up the muscles and joints, which we follow with some stretching prior to actually lifting weights. I will describe these activities before I describe the workout itself.

Key Elements of the Old Man Muscle Workout	
ELEMENT	**COMMENT**
Training frequency	2x, 3x, etc. per week
Training intensity	• Type and number of exercises amount of resistance per exercise • Number of reps • Number of sets • Pace
Training modality	Dumbbells, kettlebells, machines, etc.
Workout duration	10, 20, 30 minutes, etc.

▲ **Table 4.1** The key elements of the Old Man Muscle Workout that were considered when developing it. If we follow this kind of full body training, which emphasizes hitting all the major muscle groups via several compound lifts 3x per week, we get the most MPS every week we train. This type of training, when combined with adequate protein intake and enough rest, sleep, and recovery, will consistently stimulate maximal muscle and strength gains in our age group compared to so-called split routines and other training programs.

I wish to emphasize to readers that in developing the Old Man Muscle workout, I was interested in a quick workout that would get me finished in around ten minutes, with a brief period of warm up and dynamic stretching before I actually lift the weights, and a brief cool-down and some stretching after. Total time for everything is well under 30 minutes. This approach also meant not spending much rest time between sets, typically about 30 seconds. For some, this pace may be too intense at first, and of course it is permissible to slow things down a bit. It is very important to exercise at a pace that is challenging yet not unduly uncomfortable or stressful. However, I was able to stick to this routine for the months of year one, and achieved results safely and steadily, without overtraining and without significant injury. This is our goal. And, I was able to significantly increase the amount of weight I lifted by year's end and change my strength and physique to a degree that was more than I had expected with such a brief workout and being an untrained sixty-three to sixty-four year old. It demonstrated to me that "the old body" can be responsive to this particular training regimen, which not only caused me great personal satisfaction – it motivated me to want to let others know about it.

ACTIVITY:	APPROX. TIME:
decompress ▼	
warm-up ▼	→ 5 min.
dynamic stretch ▼	→ 3 min.
workout ▼	→ 10 min
cool down ▼	5 min.
stretch	

▲ **Figure 4.1** The sequence and progression from pre-workout to post-workout. Notice that although the Old Man Muscle workout takes ten minutes, extra time is required preceding and following the workout.

PRE-WORKOUT ROUTINE: WARMING UP AND STRETCHING

This point cannot be emphasized enough: DO NOT ATTEMPT TO LIFT ANY WEIGHTS until AFTER you have warmed up first, and then stretched. The reason is that an untrained older man is more likely to sustain an injury to the muscles or joints or tendons without first warming up and stretching, plus there is the potential danger of a life-threatening cardiac event if we do not warm up and stretch. Completing a pre-workout routine is not being a sissy – it is being SMART. If you launch right into lifting, your body is not really prepared for the activity. The blood circulation has not been stimulated, muscles and joints have not been adequately activated and gently stressed, and this can cause them to fail under the resistance loading (dumbbell weights in our case). If they incur damage, then you have to sit out days and possibly even weeks of workouts, and fall far behind our training schedule. An additional point is that improper training, with too much weight and at too intense a pace, could potentially initiate a serious cardiovascular event – raise the blood pressure too quickly and too soon - and the individual may suffer cardiac arrest and not be able to recover to ever workout again. As you will agree, this is not an option for us. Let's be smart and train smart, which means just a few minutes of preliminary warm up and stretching, prior to lifting.

Therefore the first order of business is to "prime the pump" prior to lifting weights with our pre-workout routine. I recommend starting with a few warm up exercises and follow these with stretching exercises that are dynamic (incorporate motion) rather than static stretching, in tune with enhancing our mobility and flexibility. This is a great way to complement our strength training workout. After all, we want to be mobile and flexible, not just strong – and injury-free.

Warm Muscles and Joints Are More Responsive To Training

So, before stretching, we have a few options regarding the warm-up. We can (1) walk outside around the house (or inside if need be) for five minutes; (2) use a treadmill for brisk walking or a stationary bike, for five minutes; or (3) perform a warm-up that is a "dress rehearsal" of sorts in which we will mimic actual workout maneuvers but without any weights. Any of these warm-up strategies can be effective – select the one that is most convenient for you. You can even mix and match: a few minutes of walking followed by a few more on the bike, or performing the dress rehearsal lifts. By going through the motions without weights, we achieve two important things: (1) we slowly and gradually increase the stress on the heart, also establishing mental blueprints for proper form for each exercise (very easily done when no weight is being lifted); and (2) we warm up our musculoskeletal system to prepare it for lifting weights. And by this I do mean that we actually elevate the body temperature and cause muscles and joints to be warmer. This enables the circulation to deliver nutrients needed by muscle during the workout at an accelerated pace, allowing the muscles and joints to be more flexible rather than cold and stiff. We must always remember that with warm muscles and joints comes better responsiveness to training.

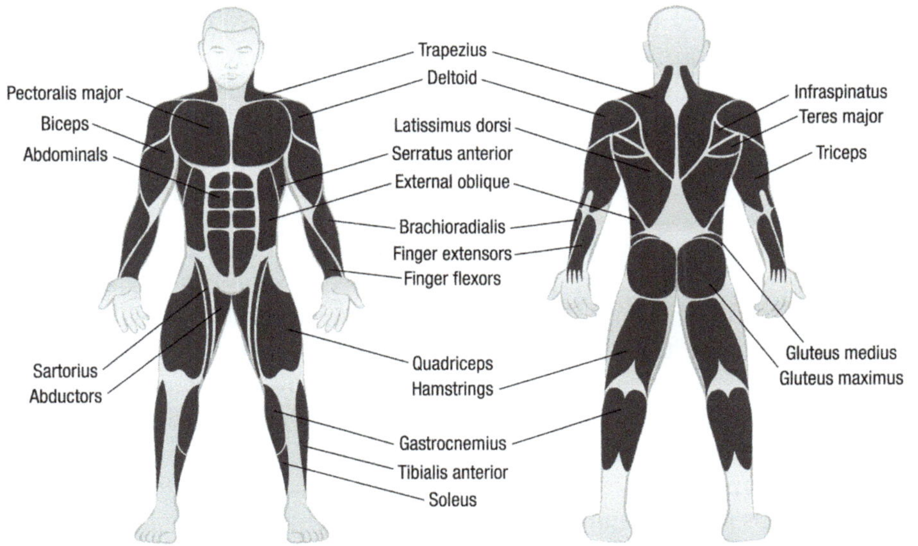

▲ **Figure 4.2** The Old Man Muscle workout is a comprehensive strength training program targeting the major muscle groups at each workout. These muscles include (roughly from bottom to top) the calf muscles (gastrocnemius or "gastrocs"), the hamstrings (behind the thigh), the quadriceps or "quads" (thigh), the glutes (gluteus maximus and medius), the abdominals and obliques (the "core"), the back including the lats (latissimus dorsi), the chest (pectoralis major), the deltoids (shoulders), the trapezius (or "traps") between shoulder and neck, and of course the biceps, triceps and minor musculature of the arms.

Warm-up examples

Before stretching, let's say that we decide to complete one set of 10 reps of several so-called "rehearsal exercises" as our warm-up. A rehearsal exercise example is performing curls without using any weights. We would get into proper body position, and imagine that we are holding the dumbbells, as we complete each set, flexing forcefully as if there were weights in our hands. We don't just swing the arms aimlessly. What this allows for is a practice of proper form, and also the exercise movement with complete range of motion engaged. The rehearsal exercise block that we can use might include the following exercises: (1) thruster, (2) curl, (3) pull-in, (4) leg press, and (5) seated leg lift. These five rehearsal exercises involve both seated (bench press, leg press and pull-in), and standing (curling, thruster) positions. For a detailed description on how to perform 1, 2, and 3, please refer to the upcoming information that explains each exercise. I will explain the leg movements in the next paragraph. Since we will complete 10 reps of each of the five rehearsal exercises (two of which require ten per side), it is a total of seventy reps of warm-up. This should only take a couple of minutes.

The leg press as a warm-up exercise is completed without any weights being used. If you did this as a weighted exercise, you would use a leg press machine. But for the rehearsal version, you simply sit upright on a bench (or chair). If you have an adjustable incline bench, you might lean back on it a little. For the leg press movement, the leg is brought in tight toward the chest, and then pushed all the way forward to complete one rep. It is similar to driving a car and slamming on the brakes. Perform ten of these per leg. You should feel the quads and hamstrings engaging, as you lead with the heels of your foot.

Regarding the seated leg lift, again as a rehearsal exercise, we use no weights. Sit on a flat bench or chair as before, but this time, keep your leg straight (unbent) facing forward, and raise it up as high as you can, then lower it back down. This may be about twelve inches or so, and you should feel a burning in the hamstrings as you perform each leg lift. It is important in this exercise to keep the leg straight and not bend the knee as you move your leg up and down. The resting leg simply remains in the neutral seated position until you switch over to exercise it. Again, performing ten of these per leg is the goal.

An alternative warm-up to the previous five exercises would be to walk briskly for about five minutes. This will get the heart pumping, the muscles and joints engaged, and the blood flowing. As you walk, count the number of steps during the five minutes. That way, you engage your brain, making it a good exercise for the mind as well as the body. Or, you could get on a treadmill for five minutes, or an exercise bike. If you have access to equipment such as battle ropes (perform the "slam" repeatedly), or medicine balls, weight sleds, or tires (or heavy boxes) for flipping, you can incorporate any of these. I guarantee that these activities will get the blood pumping and raise your body temperature in a hurry. You know when the breathing becomes labored and the sweat begins to drop, that you are doing a good job here. Do keep in mind that the key is not to move too much weight during the warm up if you are flipping tires or pushing/pulling a weighted sled – you do not want to tire the muscles since you need them to be fresh for the demands of the workout.

Dynamic Stretching

Dynamic stretching is the perfect follow-up to our warm-up, and involves body movement, which is in contrast to static stretching, which only requires holding a position. Dynamic stretch exercises are therefore good mobility exercises. There are five dynamic stretch exercises that I recommend, from among the many worthy options that exist. They are: (1) the inchworm, (2) the hamstring reach, (3) the can opener, (4) dead shoulder circles, and (5) calf raises. For each dynamic stretch, we complete either ten repetitions, or in the case of the can opener and the dead shoulder circles, twenty repetitions, ten with the left arm leading, followed by ten with the right arm. In all, this provides seventy high quality stretches to the body's soft tissue (muscles, tendons, ligaments) and joints prior to our workout – which is highly desirable.

The Inchworm

The **inchworm** stretch is effective in loosening up the back and also the hamstrings, and allows the arms, back, shoulders and the core to get involved too. To begin, stand up straight and tall. Then, reach down to the ground almost as if to touch your toes, but keep going forward, walking on your hands until you are in almost a push-up position. You will probably need to bend your knees slightly as you move forward and down. Then go in reverse back up to the standing position. Do try to keep the knees from bending much in this exercise, and you will feel it in the hamstrings (the back part of your legs), which is the stretch you are after. If at first you cannot perform the inchworm as outlined here, you might do an easier modification of it reaching down to the seat of a chair to get a back and hamstring stretch, and straightening back up.

The Hamstring Reach

The next stretch is the **hamstring reach**, in which we first spread our legs apart this way – one foot and leg pushed out forward and knee straight (not bent), and the other leg pointing backward, also not bent, keeping your upper body straight upright and legs in a V-shape. For the rest of the exercise, get your arms out in front of you (at shoulder level) as if you were trying to push against an invisible wall that is closing in on you. Without allowing the knees to bend, lean and reach forward as much as possible (it will only be a few inches) and then return to the pre-reach position with your arms ready to push again, forward and back, repeat so as to do this five times on each leg. The hamstring reach provides forward flexion of the arms, shoulders, spine, and of course your hamstrings will get a workout too.

The Can Opener

Next we come to what is called the **can opener**, and it serves to loosen up the upper body. You remember the old dance craze from 1960, the "Twist" – well, this is not exactly the same, but starts somewhat similar. We want our legs apart, and to rotate at the hips side to side as we allow our arms to swing loose. So get that going. Next, as you keep in motion, think of a baseball pitcher. We raise the right arm up, and throw an imaginary ball as we continue to "twist". Our right arm will travel down across the body to the left side, while the other arm "goes along for the ride". This gives a dynamic stretch to the shoulder and the chest of the upper body. We need to do 10 of these per side (per arm, really). The can opener gets your body in motion and warms up your joints and muscles in preparation for lifting.

Dead Shoulder Circle

The next stretch exercise for our pre-workout mobility training is called the **dead shoulder circle**. These will loosen up the shoulders and get the circulation going in the upper body so that we can do our bench pressing, curling, and so forth without injury when we get the weights out. Dead shoulder circles require a hip hinge to get into position, and a circling motion of each arm individually, hanging limp. To do the shoulder circles, we start with one arm, let's say the right one, and with our legs spread widely apart, we lean forward and down, allowing the arm to dangle loosely. Keep it loose and try to draw 10 big circles on the floor without actually touching the floor with your hand and fingers. When finished, repeat with the other arm.

Calf Raise

The last exercise for our pre-workout mobility training is the **calf raise**. Stand with your feet hip-width apart. You can use something to balance, such as facing a wall and keeping the fingers of both hands in contact with it. Rise up onto the balls of your feet, lifting your heels off the floor while keeping your toes on the floor and your knees straight. You should raise your heels high enough that you feel a strong tightening of the muscles in the back of your leg (the gastrocs, or calf muscles) without any pain or cramping. By the way, experts suggest doing this one at a slow cadence using a 2-1-2-1 approach. This translates to a two-second initial calf stretch with legs held straight before raising, a one-second quick contraction up, a two-second isometric hold at the top of the range of motion, and then a one-second eccentric motion back down. Toe raises are best tackled while leaning against a support, such as a wall. For added stretch, you may use a step a few inches off the ground.

Done – that's a wrap for stretching! We are now ready to work out with weights. But first it is critical that we learn from the very beginning the correct way to perform each of the workout exercises. This prevents bad habits from creeping in, and more importantly, helps reduce our risk of injury. Bad habits include deviations in form such as improper foot and body stance and posture, lunging and lurching to one side, or rocking the body backward to make up for weakness in one group of muscles causing other muscles to attempt to pick up the slack. This is where having a personal trainer comes in handy. The trainer can spot and correct errors in technique. Many injuries occur when an exercise is not being properly performed due to such bad habits. Furthermore, improper body postures and movements place an abnormal amount of stress on a particular muscle, tendon, or joint, which then can become damaged. A strain, sprain, or even tear can be the result. Proper form ensures this will not happen, as long as the dumbbell weight (resistance load) is correctly selected and is not excessive.

Another point I need to emphasize is the importance of proper breathing during any warming up, stretching, and of course each exercise. It is a big no-no to strain and push and otherwise move or stretch while holding our breath. Why? This creates immediate and potentially dangerous increases in blood pressure and places a strain on the heart muscle. By contrast, when we breathe, this stress on the heart is not created. So how should we breathe? Try to remember to breathe out (think of a "puff" exhale sound) during the push part of the lift, for example if it is the bench press, as you push the weights up, breath out, and breathe in as you lower the weights, and repeat. Always try to breathe in a smooth rhythm, rather than in a "herky-jerky" manner. Another example when curling, breathe out as the arms lift the weights up, and breathe in as you lower your arms. With a little practice, you will start doing the proper breathing automatically for each warm-up and exercise. Normal or slightly slower, deeper breaths work well for the stretches.

THE OLD MAN MUSCLE WORKOUT – Background

Ideally, your Old Man Muscle workout needs to be set to occur on three non-consecutive days of the week. For example, it could take place M-W-F, or T-Th-Sat. Having two weekend days off is great, both from the standpoint of providing adequate rest/recovery, and because many folks are busy with life on the weekends. Let me also point out that fellas who are really up there in years, or who have certain extenuating circumstances, could work two days of the week instead of three, and still gain benefit. For example – train on Monday and on Friday only. What has been effective in my experience is to combine all major muscle groups into a single day's workout, rather than to schedule upper body one day, lower body another, and repeat. The latter strategy, called a "split routine", can be fine but ours takes less time, hits all the major muscle groups in a single day, and will do so safely. Another possibility is that training whole body is the anabolic way to go, whereas I have heard that split routines may be a bit catabolic, the way that endurance (aerobic) training can be catabolic rather than anabolic.

Advantages of the Old Man Muscle Workout
• Can be completed in ten minutes.
• Minimizes the risk of injury and boredom.
• Correctly requires warm-up and stretching activities before lifting.
• In conjunction with sound dietary practices, helps with weight loss.
• Uses dumbbells as free weights for superior muscle fiber recruitment.
• Age-appropriate in terms of resistance loads chosen and workout intensity.
• Works both upper and lower body to achieve a balanced and stronger overall physique.
• Safe and effective format that includes compound exercises as well as isolation exercises.
• Will result in stimulation of anabolic hormones and cell molecules to restore health and vitality.
• Employs periodization in which there is progressive overload pre-programmed into the workout.

▲ **Table 4.2** Advantages of the Old Man Muscle workout.

In order to be age-appropriate and to avoid injury and overtraining, we will be conservative regarding the amount of weight we lift during the initial phases of training, and also thereafter during the bulk of the first year. At 4 months into training, you should feel and see visual evidence of muscle development and can safely start increasing the weight, although only slightly at first. The strategy will be to build up our strength and muscle size gradually – which means SAFELY – to a point at which we have developed the all-important mind: muscle connection (a form of neuromuscular adaptation in which new motor units – nerves connected to muscle fibers - are established), which in a simple sense means we mentally instruct various body muscles (arms, chest, back, legs) to flex/contract. This is in contrast to when we begin the training, because at the beginning our muscle development will be poor by comparison, and the amount of connections between the muscles we have developed and the brain will be minimal. How many untrained old men out there can make the pecs flex at the beginning stage of strength training? Probably none. But, we should be able to develop this ability, and others, in time.

The exciting fact that will become evident in your training life will be that if you consistently put in the time to train, and use your mind as well as your muscles, you WILL get results. And that brings me to another important side point – that all of our hard work training in the gym will be FOR NOTHING if we do not combine it with these three important practices: (1) a healthy diet that will help us achieve and maintain a healthy body weight – which for many, will require losing weight; (2) adequate rest (programming an off day between workouts) for our muscles and for our nervous system to recover (we must realize that a key player in muscle and strength development is the nervous system, which works just as hard for us as do our muscles); and (3) adequate sleep each day – this is when testosterone and growth hormone are secreted to anabolically build our stronger and more muscular physiques.

THE OLD MAN MUSCLE WORKOUT – Phase One

During Phase One (months 1-4) of our Old Man Muscle workout, we have a special first two weeks of training – remember, three sessions per week - to complete during the first month of lifting. These first six workouts will involve only light lifting, serving as "get the body acquainted to weight training" workouts. This will help establish mental engrams during which time we learn the proper body positions and lift patterns that will serve us so well as we progress in our training. As always, we will begin these initial sessions with first our warm-up followed by dynamic stretches, as outlined previously. After these are done, we will immediately complete one set of reps of the four rehearsal exercises mentioned previously. Taken together with our dynamic stretching, we will have performed well over one hundred exercises in our warming up and stretching. We are now ready to lift!

For the first two weeks, and ONLY the first two weeks, we will follow our warm-up and stretching with very manageable dumbbell lifting exercises. The load will be intentionally light, at about 50% 1-RM. This means at half of one "repetition maximum". To know approximately what this is, you first need access to dumbbells of varying sizes, say, from 5 to 25 pounds in 5-lb. increments. You must determine what is the maximum dumbbell weight that you can lift for each exercise. Let's take the dumbbell curl for example. Grab a 5-lb. dumbbell in each hand, and get into position. Assuming it is easy for you to curl this weight more than once, put the fives down and try the tens, fifteens, and so forth (resting between attempts) until you hit the one that you cannot lift. Then back off by 5-lbs. This will be taken as your 1-RM for that specific exercise.

Now, let's imagine that you could not curl the 25-lb dumbbells. Go backward 5 pounds, and that means that 20 lbs. in each hand is your 1-RM for the dumbbell curl exercise. Half of that is 10 lbs., so for the first two weeks, you will only curl that amount, 10 lbs., as your 50% 1-RM. Potentially, each exercise in the workout will have a unique 1-RM for you, from which you derive the half 1-RM for the purpose of each lift. You should write these all down in a journal – use an old school technique like pad and paper, or your personal electronic device if you'd rather – whatever – but this is key information to keep track of. You want to be able to track your training volume (the sum of the number of pounds x number of sets x number of reps of each of the dumbbell exercises), to quantify strength gains made over time. When you see an increase in training volume, this means you have increased strength. This is what strength training is all about.

As a general rule, you should attempt to increase the weight lifted by between 2-5 lbs. every phase of training – which for us means every four months, which is three times during the first year. When using dumbbells, the reality is they are in increments of 5 lbs., so in this case, we might increase the number of reps if we can't quite make it to the next 5 lbs. each phase. For example, let's say that during Phase 1 of Year 1 training, our 50% of 1-RM for the single arm row starts out at 15 lbs. This means we use a 15 lb. dumbbell for perhaps 10 reps in that exercise. Starting in Phase 2 (months 5-8), we would attempt to increase the weight to 20 lbs. But if that weight feels too heavy for 10 reps, stay at 15 lbs. but increase the number of reps by 5, which means instead of 10 reps x 15 lbs., perform 15 reps x 15 lbs. That will increase your training volume for the exercise significantly: from 150 lbs. to 225 lbs. Perhaps by Phase 3, you can make the change to 20 lbs. x 10 reps, and after a while, to 20 lbs. x 15 reps. When I can perform 15 reps without total burnout on the 15th, I take that as a signal that it is time to increase the weight and lower the number of reps back down to 10
(or eight).

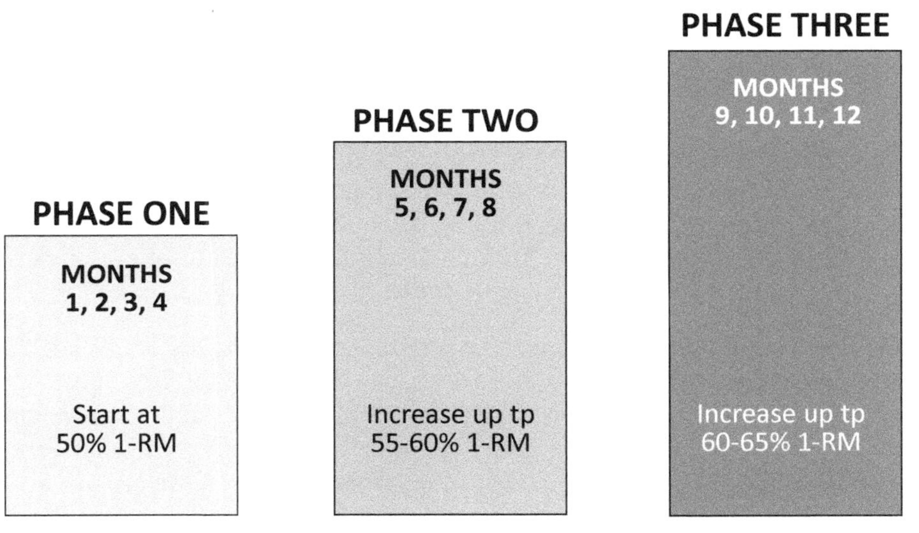

▲ **Figure 4.2** Phases of the Old Man Muscle workout. The year is divided into three phases, each of four weeks duration. The goal is to try to increase the weight lifted per exercise as you progress from one phase to the next. Or, increase the number of reps within sets if the weight cannot yet be increased. This will achieve a progressively greater training volume, and by the end of the year, an increase in the 1-RM for each of the ten exercises of the Old Man Muscle workout. This is progressive resistance training in action.

By the end of the year, you will be able to see how your 1-RM, or training volume, has increased as you gain strength and muscle. For example, if you started curling 10 lb. dumbbells in each hand, see if you can increase to 15 lbs. by month four, to 20 lbs. by month 9, and to 25 lbs. by year's end. Your 1-RM might have started at 20 lb., and by year's end, increased to 30 lb. or more. That would be real progress. Notice: if you do not make it to 25 pounds, this does not mean that you fail! You have made progress that can be seen and felt – and can built upon in **year two**. It is probably more important to show any increase than to force a number, although it is great to have a specific weight goal by year's end.

From week three onward during the remainder of month one, and during all of months two, three, and four, we will be sticking to 50% of 1-RM. I feel that this is the safest approach for untrained older men. However, we will try increasing the weight of the dumbbells by a minimum of 5 lbs. each during Phase Two (months 5-8), bringing us to a higher percentage of 1-RM, which will be similar to how the younger guys train. We will be making steady progress. And in Phase Three to finish off our first year, we will double our workout to two sets of each of the ten exercises, instead of just one, which doubles our training volume. This will also double the time required to complete the workout: increasing it from 10 to 20 minutes. But, we should still be able to complete the entire workout (warm-up, stretching, lifting) in under 30 minutes. I will now describe the workout exercises in full detail.

Now that we know what the Old Man Muscle workout entails during the first phase of our first year of training, it is time to discuss what happens during Phase Two (months 5-8) and Phase Three (months 9-12).

THE OLD MAN MUSCLE WORKOUT – Phase Two

As we already know, the Old Man Muscle workout program has built into it a progressive resistance strategy. During Phase One, we have trained ourselves to warm up and to stretch prior to each weight lifting circuit of ten specially selected exercises. Phase One by the way is a unique and special time for all of us. The fastest changes come during Phase One. It is truly the "honeymoon period" of our training when the most rapid changes can occur. For me, this was during the summer of 2017, and I realize that I cannot ever experience my Phase One ever again; it was a special time. So make the most of yours!

During Phase Two, we actually implement the progressive training aspect. What this requires is that we reassess what our 1-RM is for each of the ten exercises, now that we have been gaining strength. We can do this easily by determining which dumbbell weights for each exercise now correspond to 1-RM, and then scale back by 50%. Alternatively, we can simply try adding 5 lbs. of weight (or increase the number of reps if we are not able to jump up an additional 5 lbs.) to each exercise that we were doing during Phase One. We then continue with our stretching, warm up, and the slightly heavier lifting workout accordingly during months 5 through 8, which constitute Phase Two. You see that we are going slowly and carefully in our progression, and this is for a reason: to get results of course, but also to get them safely. We stick to one set of about 10 reps for each of the ten exercises, which keeps the workout at ten minutes. Remember: the Old Man Muscle workout is an age appropriate plan. It recognizes that at our age, we have certain limitations superimposed upon us. If you are impatient and wish to work at a more intense level, believe me, you will love Phase Three! But as it is, we must be realistic, and do things right, and this means progressing gradually, without extreme weights. We are only truly ready for Phase Three when we complete Phases One and Two.

THE OLD MAN MUSCLE WORKOUT – Phase Three

With Phase Three, which happens during months 9 through 12 of year one, we close out our first year of training using the Old Man Muscle workout. When Phase Three finally arrives, at the start of month nine, you will by now have noticed some positive changes in yourself. You are beginning to see muscular development, and strength increases, and you are becoming mind:muscle proficient during your lifting. And these things will continue, although at a relatively slower pace, as long as you keep training. This is normal, and is due to neuromuscular adaptations that occur due to training. So now that we are in better shape compared to when we started nine months ago, it is time to DOUBLE the number of sets, using the same weight ranges that were incorporated during Phase Two. This means exactly what you think it means: each of the ten exercises now must be completed in not one, but in two sets of up to twelve reps each. This will effectively double our workload, output and training volume.

A question that you may have at this point, and it is a good one, is "Should I go through one complete circuit of all ten exercises to get in the first rep for each, and then start all over again and repeat for rep number two, or should I take care of the double sets as I move through all the ten exercises in sequence?" Although theoretically Phase Three would allow for either option, I will tell you what I did, and it worked. I chose the second approach. Starting with the deadlift, I did two sets of it, before moving on to the next exercise, sequentially getting through all ten exercises that way. This brought me to the end of year one.

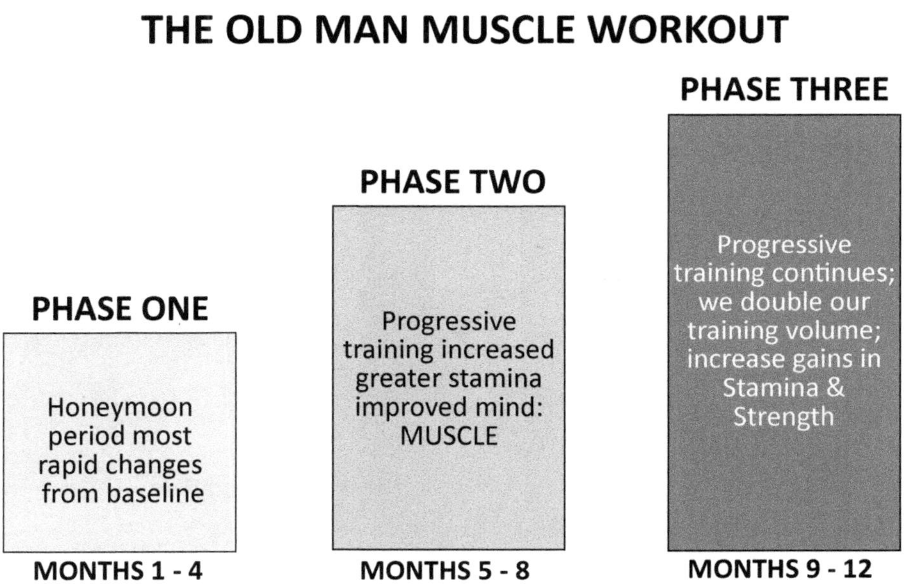

▲ **Figure 4.3** Workout summary for Phases One, Two and Three of year one of the Old Man Muscle Workout.

For all of the Old Man Muscle exercises, we will use free weights, specifically dumbbells. We will not use barbells or machines, although they certainly can have their place in a workout plan later. Interestingly, it has been asserted by experts that motor skill engrams are not established as well using exercise machines as they are with free weights/dumbbells. It is not that they fail to be established when using machines – just that dumbbells are superior in this regard.

I would add anecdotally a psychological difference - that I feel better and more satisfied with my workout when I use dumbbells. Another point regarding machines is that some tend to isolate specific muscle groups and do much of the assistance work that our supporting muscles would normally do. This means that dumbbells are superior to machines in their ability to recruit more muscles even down to the level of the muscle fibers – the target muscles plus supporting muscles - as we train. This recruitment pattern translates to improved strength gains, and is a better match to real world muscle movement than the recruitment pattern obtained through machine use.

However, to be fair to machines, I would say that they do have a significant potential to help us create a good aesthetic physique when properly used. And if a person is recovering from an injury or has significant joint pain, the machine would be a good choice until they are back to normal. Lately, new exercise machines are being designed that train multiple muscle groups simultaneously, which is positive. So please understand that I am not anti-machine. It is just hard to beat dumbbells. Among other benefits, dumbbells allow for three-dimensional movement within limited space, which equates to allowing for full range of motion. Also, dumbbells are relatively small and easy to control compared to a barbell. So it is important to have access to dumbbells of various sizes, plus a workout bench, such as your typical incline exercise bench or flat bench. In a pinch, a simple sturdy chair may do. Simple utility exercise benches are not too expensive and represent a good investment in your health, as do the dumbbells.

▲ **Figure 4.4** You will notice when training that there are two ways to grip the dumbbell with your hand. The difference in grips: palms up is called supinated, palms down on the dumbbell is called pronated.

With all of these workout exercises, proper form and execution is more important than the amount of weight you are lifting at the start. I say this because without the proper form, the target muscles will not be worked to good effect, so it is better to lift a bit lighter and more comfortably at first. If too heavy a weight is chosen causing improper posture and motion during lifting to overcompensate, an injury can occur. Lifting in order to achieve the benefits of strength training as an old man is not about impressing yourself - or anyone else. As the saying goes. "Check your ego at the door" if you are serious about weight training! Don't worry about what Joe over there is lifting – you are training for yourself and no one else.

DEVELOP THE MIND:MUSCLE CONNECTION PRACTICE MAKES PERFECT

At this point, I want to remind us about a few things regarding our muscles and how their connection to nerves (neurons) is important in our training. You see, one of the main reasons for failure to see results in spite of lifting weights for months at a time is the failure to develop the mind:muscle connection described previously. We must put the mind:muscle principle into practice in each and every one of our lifts. It goes like this. All body movement starts with the brain. If you want to move your arm to pick up a dumbbell and execute a curl, the very action is initiated by a message from the brain to the proper muscles in the arm. Remember that the brain is connected to neurons that extend out to the muscles throughout the body – the arms, legs, chest, and so forth. Thus, the brain is truly connected to all our muscles.

We have the ability to focus neural connections to specific muscle groups that we want to work, and leave out most of the others. But this won't just happen. If you don't think about performing a curl properly as you practice the movement, many other muscles besides the biceps (those in the wrists, forearms, shoulder, back, etc.) can become activated and sort of "steal the show", with the result that the biceps are cheated out of the full workout and fail to grow – even though we did our curls. To avoid this mistake in training, we need to think about and feel (by flexing it) the muscle group that is the "target" – the one we are interested in working. We activate them by this flexing, and when we begin the exercise, we try to isolate these flexed muscles as the "prime movers" of the dumbbell, rather than any of the other muscles. It really is an amazing technique to adopt, this mind:muscle thing - and will generate better results than if you do not use it. Continue to work out this way throughout all your exercises – and this will eliminate unwanted muscle activation from other muscles that we do not want to bring into the picture.

Now – what about that promised training program? It's time to examine together each of the ten Old Man Workout lifts. These are the ones that I have personally used successfully and can recommend from experience for the first year of training. By the way, I have noticed by reading comments about other exercise books that – amazingly - some authors do not include workout specifics! Here is one person's comment regarding such a book: *"I was disappointed that after he went through all the effort to determine and discuss best ways for burning fat and increasing muscle mass, the author didn't provide sample workouts to follow. It would be nice to see a recommended schedule and exercises it should include. If you want to find that out, you have to buy another book! Similar with nutrition. Laid out a few diets and what he tried along the way, but lacked detail. You would need to buy another book on whatever nutrition plan you decide on."*

Fear not – I will provide the information you need in this Old Man Muscle Handbook. And for those not having a personal trainer to assist them, I will do my best to describe each exercise in the upcoming paragraphs as if I was there with you. Here is a chart that captures all of the Old Man Muscle exercises and categorizes them plus summarizes the training volume of this workout.

OLD MAN MUSCLE 10-MINUTE DUMBBELL WORKOUT						
EXERCISE	TYPE	UPPER/LOWER	SETS	REPS	WEIGHT	OUTPUT
Dead lift	Push/Pull	Both	1	10	2x15	300
Goblet Squat	Push	Both	1	10	1x15	150
Thruster	Push	Both	1	10	2x10	200
Bench Press	Push	Upper	1	10	2x15	300
Curl	Pull	Upper	1	10	2x10	200
Triceps Kickback	Push	Upper	1	10	2x10	200
Pull-In	Pull	Upper	1	10	2x10	200
Single Arm Row	Pull	Upper	1	10	2x15	300
Trapezius Shrug	Pull	Upper	1	10	2x15	300
Swing	Pull	Both	1	10	1x15	150

▲ **Table 4.3** The complete exercise plan for the Old Man Muscle workout, year one. The ten exercises are grouped according to whether they are primarily push type or pull with respect to muscle movement, and whether they work the upper body, lower body, or both. Each exercise is performed once (one set), with 10 repetitions of each exercise within its set (in actual practice, you may choose 8, 10, or 12 reps). If the weight to be moved is listed as "2 x 15", this means two dumbbells are held, one per hand, each weighing 15 pounds. Certain exercises work both arms simultaneously, while others do so singly (see description specifics). The total output (volume) of the entire workout is the sum of each exercise's output as given in the far right column, which totals 2,300 lbs. if performed as written.

THE EXERCISES OF THE OLD MAN MUSCLE WORKOUT

Very important reminders: (1) be sure that you perform a warm-up, followed by the dynamic stretches presented, before attempting the exercise phase detailed below; (2) be sure to adequately hydrate as you progress through the workout, and at the conclusion.

1. DEADLIFT (up to 12 reps, barbell or dumbbells)

This first exercise, the **deadlift**, is probably the most difficult to explain and to execute, even with no weights. So I will be saying a lot here. The deadlift can be performed with dumbbells, which is the way we will do it at first. But perhaps later in the year, you will want to use a barbell for more weight. Also, it feels better from the ground up this way, and you really know that you are getting every major muscle involved when there is a heavy bar with some plates added sitting as a "dead weight" on the ground, waiting for you and daring you to lift it. A major challenge of this lift is to get into the proper starting position and use correct form during the lift so as to not stress your lower back, which might lead to soreness or an injury. You really have to get down into almost a squat position (if you are physically able) and stick your butt out, which means moving your hips back, and prevent any hunching or rounding of the back and internal rotation of the shoulders. The movement I am describing is actually not a true squat position, but one called a "hip hinge" – more on this in a bit. Those who perform the deadlift incorrectly often make it look like a hunched and bent-over squat – caution: not good for the back! In fact, if you have ever suffered a herniated disc or some other back injury, you should skip this exercise altogether, and focus on the squat instead.

▲ **Figure 4.5** The deadlift works the major muscle groups and joints, and is one of the key foundation strength training exercises. The upward hip hinge movement should kick in when the weights move past the shins to knee level, raising the weights high enough so that you are back in the standing, upright position. Lowering the weights down in reverse (using a downward hip hinge) will prevent you from hurting your lower back. Read the description below to gain further insight on when to engage the legs versus when to engage the hips – this distinction is crucial.

As indicated, dumbbells or a barbell can be used. I will first describe the dumbbell version, and then the barbell. Grasping one dumbbell in each hand, try to grip them and hold them as if you were holding a barbell, and then perform the exercise as described above. To get a good workout, you will need to lift a somewhat heavy dumbbell in each hand. Remember – your arms should not do the lifting, your glutes, hamstrings, and hips do, and when brought into play they can lift quite a bit compared to the arms. Notice that we keep the legs stiff (unbent) as we lower the dumbbells, making this exercise a stiff-legged deadlift. We can also try for the same leg position on the way up, but be sure if you do, to keep your back from rounding, which will strain the lower back. A neutral spine position is always recommended instead, in which the back is not arched but rather is straight, or "flat". The core area should also be called into play during the up and down of the deadlift in order to prevent any back strain. While it is true that the technique addressed above with respect to using a barbell will allow for greater weight to be lifted over time by progressively adding more plates, the dumbbell deadlift is more appropriate for our age group, as our goal is not really to progress to deadlifting our body weight.

A challenge is to develop the mind:muscle connection in such a way so that the hips become more engaged than the arms, lower back, and knees in this lift – a difficult thing to achieve without thinking about what you are doing before as well as during the lift itself. You want to imagine driving your legs through your feet into the ground before you even start the lift, as this exercise is primarily a push exercise from the bottom up; the supporting muscles like the arms only do the secondary pulling to stabilize the bar rather than to lift it. Look down at the bar as you descend. As soon as your hands drop to knee level, stick your hips back, bend your knees, and reach down to grab the bar. Slowly start the ascent, bringing the bar up past your shins and up to thigh level by driving your legs down and hips back as you stand vertically. To do this properly, you have to incorporate the hip hinge movement. You do NOT lift the bar up with the arms! Hold the bar at this top position momentarily before allowing the hips to drop, the knees to bend, and finally returning the bar to the ground, without releasing your grip. Make sure that on the descent, you stick your butt out as much as you can all the way down to avoid back strain. Repeat.

At the start of this exercise, I use a double overhand (pronated) grip, set my feet hip width apart, and my hands a few inches wider apart than my feet (don't want the hands to drag up your thighs – the bar yes, your hands, no). Keep the arms hanging down straight with tension – and don't bend the elbows. Let's go through this again, because the deadlift is a difficult lift to master correctly. As we grab the bar and get into lifting position, our shins should make contact with the bar. To avoid rounding the back, drive the chest out, and the hips down, and also activate the lats if you can. If your lats are not yet developed, no worries – we will be working on them in other exercises, and in time, they will grow. Keep the lower back flat and the butt out! Now comes the leg exercise part. The lift must start with a big push from the legs to get the bar up to your knee, followed by the hips – this is not to be an arm, lower back, or knee lift. Your legs are much stronger than your arms, and will move the weight for you without injury if you call upon them. So - up to the level of the knee use your legs; above the knee, the hips and mid to upper back engage (flip up, becoming "unbent") as you become vertical and the bar travels up to the point it needs to end up at: sitting on your thighs. The act of upward hip-hinging will make the bar come up – again, do not attempt to lift emphasizing the arms. To lower the bar to the floor, simply reverse these motions. Use the downward hip hinge to get the bar down to the level of the knee, and then let the knees bend and stick your butt back and keep your back straight, not arched – and straight down goes the bar to the floor. Due to all the hip motion required, it is not a bad idea to practice the hip hinge, the down and up movements, before actually attempting the deadlift (and even the squat).

So now, about how to correctly practice hip hinging in preparation for deadlifting. Funny thing is, folks can conceptualize a squat much easier than they can a hip hinge! But it is actually a simple move once you know what it is all about. The downward hip hinge is like a deep bow you would make were you in the presence of the king and queen. The reverse motion is used as you lift the bar above the knees.

Let's give it a try. To perform a practice hip hinge requires standing by a wall. The move requires a slow leaning forward and bending forward and down, primarily bending at the hips, which drops the torso forward with only a slight knee bend accompanying this movement rather than a major knee bend (differentiating the hip hinge from the squat). This is actually how a toddler reaches down to pick up an object – they use the hip hinge! Let's break it down. While standing about three inches away from the wall, facing away, try to tap (reach) the wall with your butt. To do this, you have to stick your butt out and hinge (bend forward) at the hips. Once you've done this from three inches, straighten up and step away from the wall another inch and repeat the same movement. Finally, step another inch forward and execute a full hip hinge. This is possible now because you are far enough away from the way that your butt will not brush up against it. You should feel a nice stretch in your hamstrings as your torso goes down, spine kept straight, never rounded or hunched. Meanwhile your knees should only have a soft bend in them. This is the motion you want to use when approaching the deadlift – like a hip hinge, not like a squat. Final word: deadlifting becomes easier, and the amount of weight we can deadlift becomes greater, when we understand how to unleash the power latent in the hips, which is what mastering the hip hinge can do for us – and this power is considerable. Our arms function merely as the means to hold and stabilize the bar, not lift the bar – and they are indeed envious of our hip strength.

2. GOBLET SQUAT (up to 12 reps, single dumbbell)

In order to perform the **goblet squat**, select a relatively heavy single dumbbell, and grip it tightly in a two-hand squeeze grip at chest level. You will increase the workout of your core if you hold the dumbbell a few inches in front of your chest rather than in contact with your chest. Drop down into a squat position, and then come back up. Repeat. This exercise trains the lower body (quads, hamstrings) and is famous for bringing us right into our natural center of gravity as we perform it every time. By squeezing rather than just holding onto the dumbbell, it also becomes an upper body exercise (chest, delts, arms become activated) as well as the core, glutes and legs. A great one.

▲ **Figure 4.6** The goblet squat provides a great workout for both the upper and lower body, and utilizes a crush grip - in effect squeezing the dumbbells with the hands. Maintaining balance may be a challenge at first!

3. THRUSTER (up to 12 reps, two dumbbells)

The dumbbell **thruster** is really a combination of a front squat and overhead press, to be properly executed in one smooth motion. It is a great compound exercise that includes a pull (initially lifting the dumbbells) and a push (overhead). With a dumbbell in each hand, raise them up to shoulder level, keeping them parallel and with the dumbbell ends facing forward. Now, drop down into a squat position. From there, push yourself into an upright position keeping the dumbbells at shoulder height, and immediately allow the momentum of the movement carry you into an explosive upward press of the dumbbells together, overhead. Progressing from the squat to this fully extended position should be performed in one fluid motion, not two separate jerky movements. Then, as you bring the dumbbells back down to shoulder height, descend into the squat position and repeat. This great exercise works just about every major muscle group and joint in the body, and is a foundation lift for developing overall strength and power. For best results, it should not be programmed at the end of the workout when the body muscles are fatigued, but rather somewhere in the middle. Also – if you have shoulder joint concerns, use an underhanded grip on the dumbbells, pushing them more to the front of your body than over your head - and you will be so much better off. I consider the dumbbell thruster as the single most important and effective exercise in the Old Man Muscle workout!

▲ **Figure 4.7** The dumbbell thruster is really a combination of a front squat and overhead press. It works just about every major muscle group and joint in the body, and is a foundation lift for developing overall strength and power.

4. BENCH PRESS (up to 12 reps, two dumbbells)

The **bench press** is one of the most famous of all weightlifting exercises, and one that most frequently is performed incorrectly. I say this because many novice lifters fail to work the chest as they should in this lift, instead focusing wrongly on using their arms. This is a "push" exercise for the chest, rather than a lift for the arms. The dumbbells should not be squeezed to death as a result – they should be gently gripped and lightly squeezed and held together during this lift. The way I suggest doing the bench press will spare your shoulders a lot of potential abuse and pain – something the bench press unfortunately has been known to cause. Rather than starting by flaring our arms out at the sides, with our elbows pointing down at shoulder height and then proceeding to push the dumbbells upward, we will instead lift with the dumbbells held touching each other (yes, lightly squeeze them together – but don't make this into an arm exercise!) starting with them positioned directly over the chest. Make certain you activate the chest by sticking it out BEFORE you prepare to push the dumbbells up (keep them squeezed together). Let the chest do the pushing – developing the mind:muscle connection. Keep the arm and shoulder engagement to a minimum. Push them up as you breathe out, and breathe in as you drop them down to your chest. Go at a relatively slow pace. As with all of our lifting, the breathing cadence is important and will become a part of your mind muscle connection, helping you perform lifts more effectively (I almost want to say more effortlessly). As you develop your pec muscles, you will see how this large muscle group has more power than your arms, and will allow you to eventually push a lot of weight off your chest in due time, that your arms could never manage. When this happens, you are becoming a master of the bench press, and you will be able to flex your pecs like the bodybuilders and wrestlers do. Expect this to happen in one year or thereabouts if you train consistently! I also need to mention that as you progress in this exercise and can lift more weight, you will reach a point where it becomes difficult by yourself – and less safe – to position yourself on the bench while you try to hold and maneuver the dumbbells as you lie down on the bench and raise them over your chest. Imagine a 40 lb. dumbbell in each hand. You really need a second person to be holding them, to hand the dumbbells to you once you are flat on the bench and ready to lift. The same person should spot you and take the dumbbells out of your hands once you are done lifting. This way you can most safely increase the weight you are bench pressing over time. Alternatively, you might train each arm independently by yourself. That way, you need no assistance. Your free arm can help position the dumbbell before the exercise, and move it to the floor after its completion.

▲ **Figure 4.8** The bench press is one of the most famous weightlifting exercises. When done correctly, it works the chest, with the arms merely serving to stabilize the dumbbells.

5. CURL (up to 12 reps each side, two dumbbells)

When considering a person to be muscular and in shape, one of the first body areas that spring to mind is the arms – especially the biceps. Large, well-defined biceps are synonymous with having a "strongman physique". However, the biceps muscles of the arms are relatively small compared to leg and back muscles, and in order to get them to grow requires some sort of isolation work. One of the more popular biceps isolation exercises is the dumbbell curl. There is the standard curl that many are familiar with, but also several variations of this. The consensus of opinion is that the **curl** is probably the finest exercise for developing the biceps, and it also works other muscles depending upon which variation you are performing. Another factor to consider when curling is will we be standing, or seated? Standing requires that our body be stabilized as we lift the weights up and down, and as such provides a bit of a workout for the core – a good thing. Sitting removes this added stress, and is an easier option for those who are just beginning a training program, or are advanced in age, are recovering from an injury, and so forth. At first in my workout I was performing a standard curl until one day I read about the curling technique favored by legendary bodybuilder Steve Reeves. This is the one I am presenting here. Stand erect with your palms facing forward (supinated), with a dumbbell in each hand. Curl the left arm upward until the biceps has reached full contraction (the concentric phase of the lift), and then twist your wrist to turn your palm down as you lower the dumbbell gradually until the biceps reaches full extension (the eccentric phase of the lift). Immediately as you are lowering your left arm, begin the same exercise with your right arm by curling your right arm upward and repeating the exercise with this arm. This is reportedly how Reeves alternated arms on this exercise - meaning, when his left bicep was contracted, the right one was extended and vice versa. The biceps as well as forearms will be developed. We wish to mimic his technique, which means we too will be constantly moving both arms – but in opposite directions, so to speak. If this sounds too confusing, you can synchronize the arms to move together in the same direction up and down instead of alternating them. By doing so, you actually cause the core muscles to engage and work more, which is not a bad thing at all.

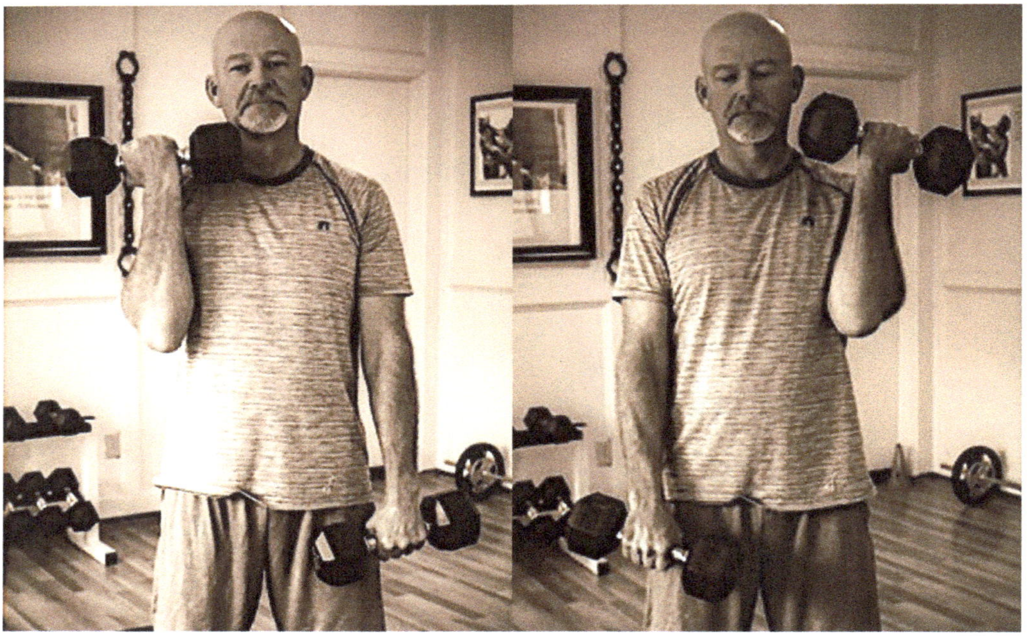

▲ **Figure 4.9** There are many variations of this exercise (standard curl, preacher curl, hammer curl, etc.). The curl can be performed standing, seated, alternating left and right, and both arms simultaneously. Shown is the alternating version that also switches the grip from pronated (on the way down – working the forearm) to supinated (on the way up, focused on the biceps).

6. TRAPEZIUS SHRUG (up to 12 reps, two dumbbells)

The **trapezius shrug** (shoulder shrug) exercise is often performed with a dumbbell in each hand with the person simply shrugging (lifting their shoulders up), but the version I am advocating requires single arm dumbbell work. For instance, start with the right arm. Grab the dumbbell and hold it at your right side, but move it out and slightly forward from your body, and have your left leg positioned backward, giving you a more stable power stance – this is the starting position. To do this exercise for full effect, shrug your shoulder and rotate your body to the outside slightly, as you pull the dumbbell with your right hand toward your right hip. The action of the shrug contracts the trapezius muscle, and the rotation ensures a fuller contraction of it compared to the standard shrug. The effective sequence to the movement is: shrug, rotate, and pull. Each time after the pull, imagine there is an invisible pulley on the dumbbell that returns it to the starting position in front of your body. Using your arms to make that happen keeps tension on the arm and on the traps, which is what you want. When finished, place the dumbbell in the other hand and repeat the exercise. In time, you may develop your mind:muscle connection enough that you can work both shoulders in an alternating or even simultaneous manner.

▲ **Figure 4.10** The standard trapezius exercise is a shrug of the shoulders, in which dumbbells weigh the arms down and provide the resistance load. The effort should be focused on the traps rather than the arms, which merely "go along for the ride". A more involved trap exercise is described above, that creates a more complete contraction of these muscles.

7. TRICEPS KICKBACK (up to 12 reps, single dumbbell each side)

Although the **triceps kickback** may not be the best exercise mechanically speaking, I used it because I could feel it affecting these muscles, and I saw growth. With proper slow, controlled technique (avoid rocking the shoulder and the elbow) you should be able to feel the triceps doing most of the work in this exercise. The technique is to stand with knees bent and lean forward slightly, with a dumbbell in each hand. Alternatively, if you lack good stability, you can rest upon a bench as shown in the photo. Keeping your back straight, bend your arm 90 degrees at the elbow to align your triceps with your back. Engage your core (squeeze it) and your triceps, and hinge at the elbow, lifting the dumbbell up and back as you try to straighten your arm behind you. Keep your triceps still and only move your elbow. Guide the weight upward until your arm is straight, pause, then lower it back to the starting position. Try not to swing your upper body and keep your spine neutral (flat back, not arched).

▲ **Figure 4.11** The triceps kickback as expected mainly works your triceps, but also strengthen the auxiliary muscles that help extend the hand.

The Only Bad Workout Is The One You Don't Do

First They Ask You Why You Do It - Then They Ask You How You Do It

8. PULL IN (Seated Row) (up to 12 reps, two dumbbells)

I think of this as the "**pull-in**" exercise, because we pull rather than push the dumbbells, as we bring them inward to the core area. This is an exercise that we complete in a sitting position. It requires that we have a dumbbell in each hand. Start by resting your arms outstretched over your legs (thighs) with the hands gripping the dumbbells "palms up" (supinated). Now, to do this exercise, elevate your arms off your thighs slightly and pull the dumbbells in toward you, so that they make contact with the sides of your abdomen just below the ribcage. For maximum effect, as this exercise is as much for the chest as it is the arms, stick out (flex, activate) the chest muscles as you pull in. Make sure to stick your elbows as far back behind your body as you can before returning to the forward position. Also, your elbows will be driving backward at an angle to move past your body – this should appear as a fluid motion. As you drive them forward to the front for the next rep, your hands will almost meet.

▲ **Figure 4.12** The pull-in is a seated dumbbell version of the seated cable row, a popular weight machine exercise. It uses an underhand grip to pull the dumbbells toward your core, and then to push them away.

A Bad Day Can Be Made Better By Working Out

Some Exercise Is Better Than None At All

9. SINGLE ARM ROW (up to 12 reps each side, single dumbbell)

For the **single arm row**, grab a dumbbell in your left hand and stand near your workout bench. Put the dumbbell down by the bench, on your left side. Place your right knee on top of the bench near the close end, and lean forward from the waist to be parallel to the floor. Then, place your right hand further up the bench for support. Next, pick up the dumbbell off the floor using your left hand and hold the weight while keeping your lower back straight. The palm of the hand should be facing your torso. You are now ready and in starting position. Pull the dumbbell straight up to the side of your chest while you breathe out, keeping your upper arm close to your side. This is an exercise for the lats and muscles of the upper back – not so much the arms. Try not to make it into an arm exercise. You can perform this one with more weight than your arms can curl, because your lats and back muscles are stronger than your arm muscles - which is how it should be. Concentrate on squeezing the back muscles once you reach the full contracted position, with your elbow pointing upward. During the movement of this exercises, only the arms, not the torso, legs, or anything else, should be moving. Lower the dumbbell straight down to the starting position as you breathe in, and repeat all over again. Don't forget to switch sides and repeat again with the other arm!

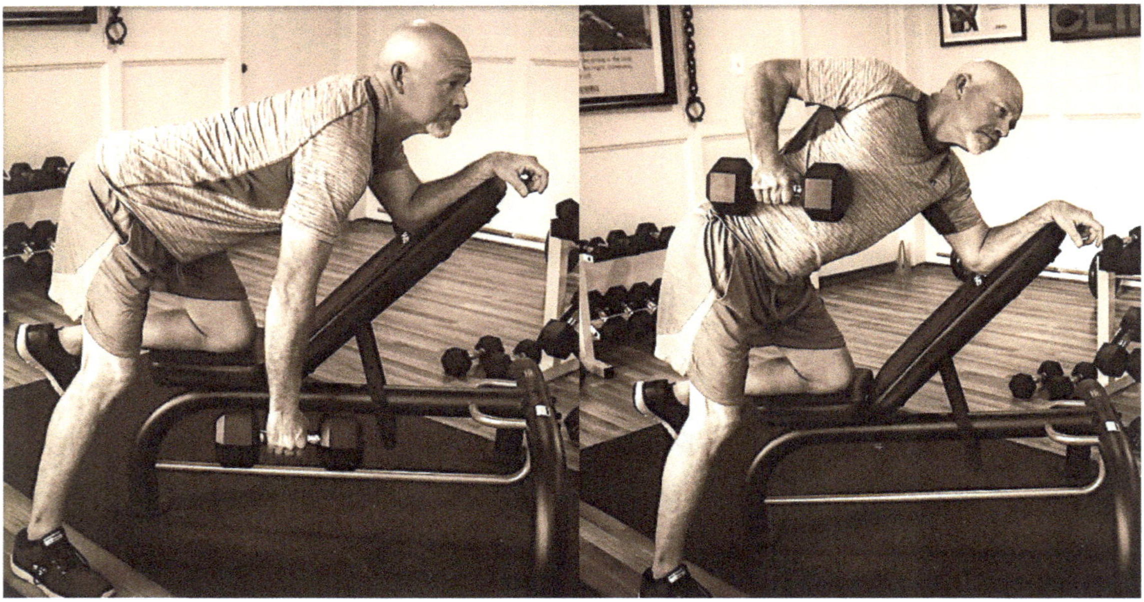

▲ **Figure 4.13** The single arm row (also known as side row, bent over row) may look like an arm exercise, but if you can concentrate on squeezing the upper back muscles together at the peak of the lift, with an assist from the lats, and only use the arm as a connector between these muscles and the dumbbell, you are on your way with this one.

10. DUMBBELL SWING (up to 12 reps, single dumbbell)

Like the goblet squat, the **dumbbell swing** is another exercise performed holding one relatively heavy dumbbell, and in this case, grasp it in the middle with one hand in front of (rather than covering) the other. Stand with feet shoulder-width apart, and without rounding your lower back, bend your knees slightly, push your hips back, raise the dumbbell and then swing it between your legs. After the weight is behind you, contract the glute muscles and push your hips forward to swing the dumbbell up to chest level. You want to use these muscles rather than your arms and back. As the dumbbell swings back up to the top position, squeeze your core, glute, and quad muscles as you allow the weight to swing down between your legs; repeat. A really great way to finish the Old Man Muscle workout.

▲ **Figure 4.14** The dumbbell swing. Keep a neutral spine by not rounding your lower back. Bend your knees slightly, push your hips back, and swing the weight between your legs, then raise it back up to the starting position. Try not to lower into a squat position, and don't raise it up over your head on the return. Breathe in as the dumbbell lowers, and breathe out on the upward swing.

Congratulations! You have just completed the ten Old Man Muscle exercises. Can you imagine how good doing this workout three times per week will be for you? I hope so, because that is our goal. This workout schedule has the ready potential over time to transform our health and physiques into something better than they are now. The key is to not simply complete each of these reps, but to make the most of each and every one by using the proper form and to concentrate as you perform them. This means to purposely activate the muscles that you are working as best as you can, to feel them contracting, before, during, and after each lift. This is part of that important mind:muscle connection you are developing, which is a conscious, deliberate thing that we are to employ during every lift of every workout.

A word of caution: the workout begins and ends before you know it – is packed with things to learn and to perform properly – it never gets old or boring, and once it's over, there's no going back to redo something. So what I am saying is, bring to each workout a bit of mental preparation and focus, so that mistakes are minimized and successful lifting is maximized.

We next discuss an important final stage of the workout: the cool-down, which begins immediately after the last exercise.

THE POST-WORKOUT COOL-DOWN

It is important to always program into your workout time the **cool-down**. At the close of the workout, when the last dumbbell swing is complete, I typically walk around the training room (or if the weather is nice, I go outside) and breathe deeply as I walk at a slow pace. In a few moments, my heart rate and breathing are back to resting levels, and I am ready to drink down my post-workout **whey protein** beverage. I typically feel a combination of an exhilarating muscle "pump" in my arms and chest, as well as fatigue, and a sense of overall well-being and accomplishment. And I give thanks for having completed another workout successfully, and without injury, at my age.

Note: if a proper cool-down is not performed, there is an increased risk of suffering **SCA**, meaning a sudden cardiac arrest. The heart rate MUST be allowed to slow back down before we resume normal activities - otherwise we are asking for trouble. Just as when we warm up (recall we needed to make sure that heart rate was increased slowly), when we cool down we must allow it to decrease slowly, as this minimizes unwanted stress on the heart. Fortunately, such stress - and a possible cardiac arrest - can be completely prevented by cooling down adequately. And the workout, when we do everything properly with the right training intensity, warm-up and cool-down, will actually make the heart larger and more powerful (being a muscle, it benefits from the training too).

THE PROPOSED BENEFITS OF A COOL-DOWN:
• Reduce heart and breathing rates
• Gradually cool body and muscle temperature
• Return muscles to their optimal length-tension relationships
• Prevent venous pooling of blood in the lower extremities
• Restore physiologic systems close to baseline

▲ **Table 4.4** A cool-down provides the body with a smooth transition from exercise back to a steady state of rest. The cool-down prevents venous pooling of blood in the lower extremities, which may cause dizziness or possible fainting if blood is not pumped to the brain. The cool-down restores various physiological systems, including the cardiovascular system, to baseline and alleviates temporary high blood pressure and elevated heart rate.

POST-EXERCISE STRETCHES

After the cool-down comes a few quick minutes of static stretching. We have learned that there is a proper time for dynamic stretching (before lifting) in order to increase body temperature, enhance joint flexibility and muscle elasticity, to prepare the body for action.

Similarly, there is a right time to perform static stretching – and it is after the lifting is over. **Static stretching** refers to holding a stretch in one position without movement, typically between 15 and 30 seconds. Be sure to properly hydrate afterward.

Post-Exercise Static Stretching

Figure 4.15 When we stretch our muscles after a workout, we are helping to give the body a jump-start on recovery, while at the same time boosting the flexibility of our joints, and releasing stress and tension. In addition, researchers now believe that the acute inflammation produced by a workout can trigger release of what are called SPMs (specialized pro-resolving mediators) that act to resolve inflammation. This serves as evidence of interaction between the immune system and the musculoskeletal system, and that stretching is an anti-inflammatory activity.

Back: Get down on your hands and knees with your hands shoulder-width apart, knees hip-width apart, abs engaged, and your back in a flat, not hunched position. This is called the "table" position. Pull in your abs as if pulling your belly button into your spine while allowing your head and neck to drop down between your arms. Hold this position for the duration of your stretch.

Chest & Biceps: While standing, reach both arms (arms straight, thumbs down) behind your back. Interlock your fingers, and with arms straight, lift arms up behind you while keeping your back straight and your shoulders down away from your ears. Hold this position for the duration of your stretch.

Hamstrings: Sit on the floor or on a bench with your legs straight out in front. Slowly lean and reach forward as far as you can, trying to touch your toes (left hand to left foot, right hand to right foot, or both together as in the figure). When you achieve the maximum reach, hold that stretch position for 15-30 seconds.

Quads: Approach a wall, standing tall, and hold onto the wall for balance with your free hand. Start out with feet parallel, and bend your left knee so that your left foot raises up behind you. Reach back and grab your left foot in your left hand, keeping your thighs lined up next to each other. Try to force your Hold this position for the duration of the stretch. Repeat for right side.

Triceps: Standing tall, place your left elbow in you right hand, reaching your left arm overhead. Try to place the palm of your left hand toward the center of your back. Keep your shoulder relaxed and away from your ears, and reach down as far as you can with your left fingertips. Hold this stretch position. Repeat with the right elbow being held by your left hand, and the right hand fingers positioned down your back.

Total body: Lie flat on your back on the floor (on a mat). Stretch your arms parallel above your head, holding them as high as possible as if a bank robber just told you, "stick 'em up!". Your legs should be together with your toes extended as far forward as possible. Hold this position for your total body stretch – what a relief it will provide after a nice, challenging workout.

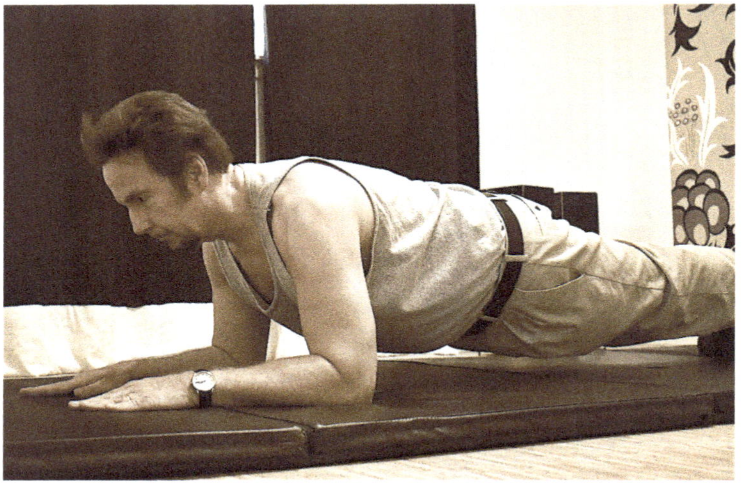

▲ **Figure 4.16** Getting into a forearm plank position and holding it for 20-30 seconds is a good end of workout strategy - it really works all the muscles of the core. Keep your back as neutral (flat) as possible – and be sure to breathe! As you become more fit, see if you can gradually increase up to 60 seconds, which would be outstanding.

Now that we have performed our static stretches, we have fully completed all of the Old Man Muscle workout components: warm-up, dynamic stretching, exercises, cool-down, and static stretching. We have achieved several things, among them: (1) elevated our heart rate, which provides not only for aerobic conditioning but also fires up cell signaling of the kind that supports muscle growth and a variety of benefits to metabolism, immune system health, and more; (2) stimulated the release of BDNF and brain endorphins, the latter providing a positive mental feeling of mild euphoria; (3) improved our balance, flexibility and mobility; and (4) completed a round of functional strength training that will allow for improved performance of our ADLs. Now is the time to have our protein beverage, and enjoy a time of rest and reflection on all that we just accomplished.

ADDITIONAL COMMENTS

Now that I have presented the Old Man Muscle Workout to you, I have a few additional things that I wish to say. First, I want to repeat what I have stated in earlier chapters that there are many different approaches to strength training and many different programs out there that are meritorious. So I don't want anyone to get the idea that I am dismissing or disparaging some other workout plans and strategies already available. What I do wish to point out, however, is that most of the time, such workouts are designed for the young, not for old-timers like us. The Old Man Muscle workout was designed by me to fill that void.

The exercises that I have presented have years and years of demonstrated success in terms of being outstanding for strength training. The sequence of performing them, as well as exercise intensity, exercise frequency, and exercise variety have all been considered by me in developing the Old Man Muscle workout. It is critical that all of these components be viewed through an age-appropriate lens – meaning, we cannot expect to work out with the frequency and intensity of a young man, at least at the start – the consequences could be a disaster. Using the appropriate dumbbell weights, our workout should have the user starting to feel results in as little as two weeks, and see results in 4-6 months, if not sooner.

Naturally, it is advisable to change the workout periodically and to gradually ramp up the intensity factor, gently at first. This can easily be accomplished by (a) increasing the resistance, (b) increasing the number of sets, (c) adding to the number of exercises, (d) reducing the time between sets, or a combination of these. And, you noticed that our plan is to start with ten minutes of training, three times per week. No, that was not a typo. As you can see, this is a very modest time commitment that should allow anyone who wishes to gain strength and muscle mass to use the Old Man Muscle workout.

Another key point is that since the workout will be over before we know it, it becomes so important during the ten minutes to concentrate on the job at hand and do the exercises as correctly as possible, with little rest in between exercises. In part, what this means is to think about each exercise BEFORE attempting it, and to mentally engage the muscle groups that will be worked. It is a great idea to "activate" those specific muscles through a sustained (isometric) flex for several seconds as we go into the exercise, using them to complete the set. In this way we will gain the benefit of the lifting by causing more micro-tears in our muscles, which leads to better muscle remodeling and gains in strength. We will also be establishing the important mental engrams that are required for successful repetition of the exercises as we continue them week by week.

I can say from experience that on more than one occasion after my workout ended and I got back to my day, I had wished I could re-do a few things because I failed to concentrate and perform them as well as I could have. And this may happen to you. Although it is disappointing when it happens, just regroup and do better next time. What happened to me was that I worked hard, but some of it was wasted effort. I will provide a quick example. The exercise I call "the pull in" works the back, arms, and chest. It is a wonderful upper-body exercise using dumbbells. If I wish to perform the exercise in a manner to lift the most weight, I must activate and involve my largest muscle groups (back and chest) more than the smaller ones (arms). When I do this exercise, if I take the trouble to learn how to "feel" these larger muscles working and bring them into focus as I lift, the pull in becomes easier, and I can lift more weight, than if I try to do everything with my arms, because the larger muscles can do more work than the smaller ones. But sometimes I don't concentrate and I make this into more of an arm exercise. This would be okay if my goal with the exercise was simply to build my arms. But such a deviation will not provide me with the gains in strength and in muscle hypertrophy and fat burning that involving the larger muscles does – so I would miss out. Bottom line: we need to pay attention to what we are doing as we lift – and make the most out of the brief time we spend lifting. Only then will all the hard work pay off.

And here's something else that may not be obvious to an old man beginner. As we know, a 50-lb. dumbbell is pretty intimidating if you are not one who has worked out and strength trained previously. If the average sixty-year-old man tries to pick it up with one hand and curl it or press it overhead – good luck! There will probably be a lot of grunting and groaning and then possibly a yell when an injury is sustained. Not good. But that same person could probably be handed that 50-lb. dumbbell, and they could eccentrically curl it, or lower it from a shoulder-level starting position. By lifting "on the way down" only, which is the eccentric rather than the concentric phase, their arm muscles could possibly handle the 50 lbs. without injury, and their arms would grow in size and strength over time if such an exercise was regularly used. This is the basis of eccentric training.

The real adventure of embarking on a strength training program at our age is to see just what kinds of results we can achieve, for instance with respect to increasing strength and building muscle, starting in year one, and to move on from there. No two persons will have exactly the same degree of result, because everyone is unique genetically, anatomically, metabolically, physiologically and psychologically. And each individual's body will respond slightly differently to strength training based upon these and a variety of other potential factors. But the human body will respond overall the same way for all of us, in that when we stress our muscles by lifting our dumbbells, a "switch' will be turned on that leads to protein synthesis and muscle growth. We will be seeking to activate this switch at all the right times, and to allow our diet and rest/sleep periods to maximize the gains that we can achieve through training.

WHAT HAPPENS IN YEAR TWO?

The information I have provided is enough to bring an older man to an improved state of conditioning, flexibility, mobility, muscle, and strength in one year's time. What happens in year two? Since the second year initiates the next stage in our Old Man Muscle training, each person needs to first evaluate their progress at the end of the first year in order to devise the right plan. This is made simple if you have been working with a trainer, but is still possible (though somewhat more challenging) if you are working on your own. The plan will take into account the same key parameters that framed our first year training protocol, namely: training frequency (3x per week), training intensity (type and number of exercises; amount of resistance per exercise; number of reps, number of sets, pace), training type (dumbbells, cables, etc.), and time (20 minute workout, 30 minutes, etc.)

▲ **Figure 4.17** If you have access to overhead cables (here I am using suspended "battle ropes"), you can perform a multitude of good, basic body weight exercises. Here I am lowering myself into a full squat position. If you have bad knees, a barbell or dumbbell squat may be out of the question. In such a case, you may be surprised to find that this variation takes the pressure off them nicely. This exercise is quite effective at working the quads and especially the glutes. These muscles need to be strong in order for us to get up and down from lying and sitting positions.

Yes, we need to be mindful after the first year of training to evaluate our progress – not only in terms of the muscle growth and training volume we achieve, but the status of our joints, our mobility, and flexibility. We may need to swap out several exercises if there is significant joint or back pain, for instance. This means that we must evaluate, adapt and adjust our workout as we age or deals with pain issues. One strategy might be to change several exercises from dumbbells to being cable-based, or to perform more exercises from a seated rather than standing position. At some point when we reach significantly advanced age we may need to lift less heavy in general, also performing fewer reps and/or sets using less weight. But no matter how old we become and how much (or how little) we can lift, it will always be the case that some exercise is better than none, and that it is just not worth trying to lift too heavy when we are really up there in years because we will destroy our joints and risk injury to muscle and other soft tissue. The goal is to maintain healthy joints and muscles and be as pain-free and injury-free as possible as we continue to age and continue to train as old men beyond the first year.

I certainly don't mind sharing with you how I have been training during my second year. It is based upon a concept that I mentioned in the previous chapter – asking you to look it up. I am talking about DUP, or daily undulating periodization. The way I have applied this concept is to vary my weekly workouts in terms of altering the intensity and training volume within each week. I accomplish this by designating a moderately heavy day, a heavy day, and a light day, making sure to work out three days per week. Also, I have been experimenting with new exercises that involve the use of barbells, kettlebells, and cable machines as well as the balance ball.

YEAR TWO OLD MAN MUSCLE WORKOUT		
MODERATE DAY (Monday)	**HEAVY DAY** (Wednesday)	**LIGHT DAY** (Friday)
10 exercises 2-3 sets 60-70% 1-RM	10 exercises 2 sets >70% 1-RM	10 exercises 3 sets <60% 1-RM

▲ **Table 4.5** Year Two of my Old Man Muscle workout includes one heavy training day, one moderate day, and one light day of training each week, followed by a weekend of rest. In addition to what you see in the table regarding the number of exercises, sets, and weight (as % 1-RM), I am experimenting with the number of reps per set, and have felt good when I perform "drop sets" with respect not to weight decrease but number of repetitions. On moderate and light days having 3 sets, the first set has ten reps, the second has eight, and the third has six; they are performed sequentially with no more than 30 seconds of rest between each set.

All in all, year two promises to build solidly upon the foundation established during year one, providing for even more gains in flexibility, strength, and muscle – exciting indeed, especially considering that we will be a year older, and in my case, approaching age seventy.

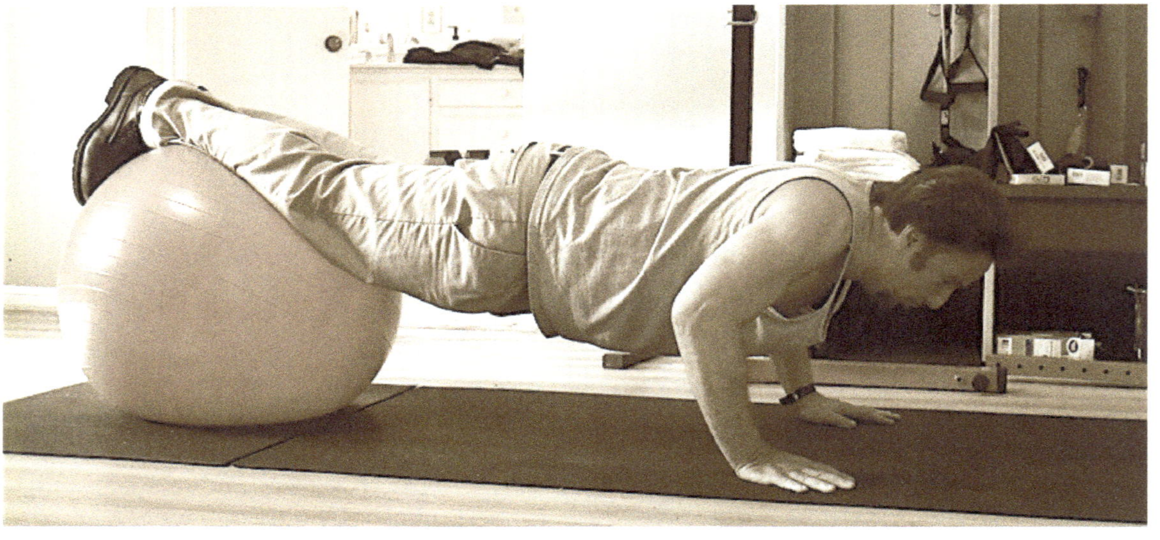

▲ **Figure 4.18** A "balance ball push-up" is a great body weight exercise to incorporate into your workout routine. Watch out - the ball will try to move out from under you, which requires concentration and core strength to maintain position. The push-up itself works the chest muscles; be sure to keep the back as straight as possible.

Ric Drasin and Doug Brignole, two older men strength trainers and celebrity bodybuilders whom I respect and admire, recommend that there needs to be some consideration regarding the way we lift as we progress from year to year in our training. Realizing that every person and every situation will be slightly different, in general we cannot continue to train the same way when we are 60 compared to when we were 50, or when we are 70 compared to when we were 60, and so on. Having said that, it may be possible for a truly dedicated individual who has everything going for him (robust health, no injuries, good genetics and so forth) to improve decade by decade up to a point (say between 50, 60, and 70) – my, that would be most impressive.

RECAP AND WHAT'S AHEAD

In this chapter, we have established the rationale for the Old Man Muscle workout as an age-appropriate progressive resistance program, and have described in detail not only the ten workout exercises themselves, but also the preparatory warm-up exercises and stretches that are part of the training program. As we train and become proficient in executing this workout, it should take only ten minutes to complete the exercise portion of the workout. We should advance gradually in our training in order to properly learn each exercise technique. Towards the end of year one, I doubled the number of sets (from one to two per exercise), which increased the workout time to twenty minutes – still relatively quick. Importantly, it also doubled my training volume, which is desirable. I had built enough strength that I could handle the increase – and it felt good. I could see and feel myself improving noticeably during this time – I hope that you will too. Please realize that the weights, outputs, and volumes presented in Table 4.3 were examples only. Each person would need to determine their 1-RM as described and then begin with a weight that is appropriate. The next chapter will focus on lifestyle factors related to the food choices we make, whether or not to take any nutritional supplements, and the importance of other lifestyle behaviors as we train.

Bryant CX, Green DJ, eds. 2010. Essentials of exercise science for fitness professionals. American Council on Exercise.

Bryant CX, Green DJ, Jo S, eds. 2014. ACE personal trainer manual. 5th Edition, American Council on Exercise.

Baechle T, Westcott W. 2010. Strength Training Older Adults. 2nd Edition, Human Kinetics.

Berrueta L, Serhan CN, et al. 2016. Stretching impacts inflammation resolution in connective tissue. J Cell Physiol. 231(7):1621-1627.

Bompa TO, Haff GG. 2009. Periodization: Theory and Methodology of Training. Accessed at: http://slovakiaclimbing.sk/wp-content/uploads/2018/06/periodizacia_kniha.pdf

Borland B. (Undated). Bodybuilding basic training.

Campos, GE et al. 2002. Muscular Adaptations in Response to Three Different Resistance-Training Regimens: Specificity of Repetition Maximum Training Zones. European Journal of Applied Physiology 88 (1-2):50-6.

Cooper aerobic-strength balance. Accessed at: https://cooperaerobics.com/Health-Tips/Fitness-Files/The-Aerobic-Strength-Balance.aspx

Cousins SO. 1998. Exercise, aging & health. Taylor & Francis.

Fletcher GF, Ades PA, Kligfield P, et al. 2013. Exercise Standards for Testing and Training: A Scientific Statement From the American Heart Association. Circulation 128:873-934.

Hatfield FC. 2019. Fitness: The Complete Guide. International Sports Sciences Association.

Jo S, Bryant CX, Green DJ, eds. 2014. ACE senior fitness manual. American Council on Exercise.

Kamen G. 2001. Foundations of Exercise Science. Lippincott, Williams & Wilkins.

Kenney WL, Willmore JH, Costill DL. 2012. Physiology of Sport and Exercise. Chapter 1: Structure and Function of Exercising Muscle. PDF version accessed at: http://medfile.ir/medstudents%20files/Learning/Physiology/Physiology%20of%20Sport%20and%20Exercise%20With%20Web%20Study%20Guide%205th%20Edition%20(www.medstudents.ir).pdf

Kreider RB. 2019. Essentials of exercise & sport nutrition: science to practice. Lulu Publishing Services.

Marijon E, Uy-Evanado A, Reinier K et al. 2015. Sudden cardiac arrest during sports activity in middle age. Circulation. 131:1384-91.

Marx JO et al. 2001. Low-volume circuit versus high-volume periodized resistance training in women. Medicine & Science in Sports & Exercise, 33:635–643.

Matthews M. 2014. Bigger Leaner Stronger. 2nd Edition, Oculus Publishing.

Minson M. 2012. Prepare to defend yourself… how to age gracefully & escape with your dignity. Texas A&M University Press.

Netreba A et al. 2007. Physiological Effects of Using the Low Intensity Strength Training without Relaxation in Single-Joint and Multi-Joint Movements. Ross Fiziol Zh Im I M Sechenova. 93 (1):27-38.

Potteiger JA. 2011. ACSM's Introduction to Exercise Science. Lippincott, Williams & Wilkins.

Reynolds G. 2012. The first 20 minutes. Icon Books, LTD.

Rikli RE, Jones CJ. 2013. Senior fitness test manual. 2nd Edition, Human Kinetics.

Serhan CN. 2014. Pro-resolving lipid mediators are leads for resolution physiology. Nature. 510(7503):92-101.

Time Magazine. 2017. The science of exercise.

Westcott W, Baechle T. 2015. Strength Training past 50. 3rd Edition, Human Kinetics.

OLD MAN MUSCLE **CHAPTER 5**

THE IMPORTANCE OF LIFESTYLE

Introduction

Each day we are active to a greater or lesser degree doing certain things. These activities range from consuming foods and beverages to walking, talking, and sleeping. How we live our lives day in and day out over an extended period of time (our **lifestyle**) has a cumulative effect on us in terms of health, functioning, and mental outlook. An unhealthy (or "bad") lifestyle results when a person engages in activities and behaviors that are detrimental to health or that otherwise increase the risk of shortening one's years on this earth. Such activities include eating poorly, not getting enough rest and sleep, drinking alcohol excessively, using drugs, smoking, and not walking or exercising regularly. Some of the hallmarks of an unhealthy lifestyle may include low energy levels, poor concentration, disease conditions, and not having a healthy weight – all markers of being metabolically unhealthy.

On the other hand, a healthy (or "good") lifestyle – one that Old Man Muscle readers ought to aspire to – is much more likely to help the person stay in good overall health and reach their maximum lifespan. By eating and drinking in a healthy, nutritious manner, getting adequate sleep each day, plus exercising regularly and avoiding tobacco products, drugs, and alcohol, we stand the best chance to reduce our risk of disease and going to an early grave.

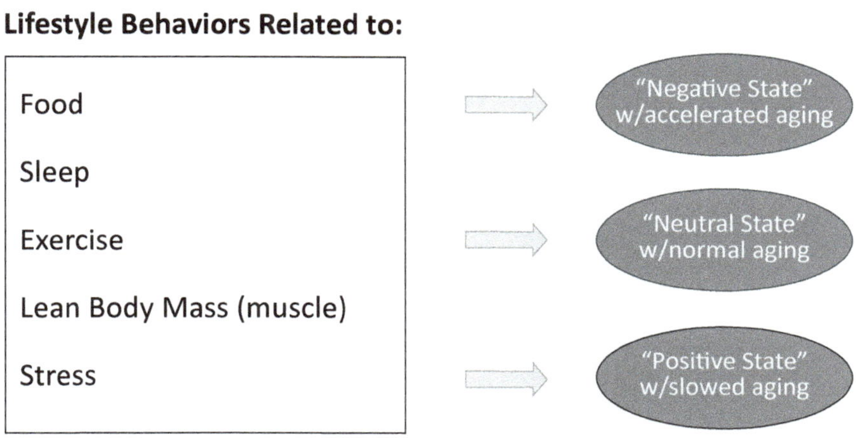

▲ **Figure 5.1** Why is lifestyle important? The relative contributions of the five lifestyle variables on the left will affect whether we will be in a negative, neutral, or positive state of healthy aging. Food is meant to include solid food, water, and certain supplements.

A healthy lifestyle involves the adoption of certain routine activities and practices that have the potential to not only help us be healthy and live longer, but also contribute to having more energy, more strength, better mental focus, more self-confidence and self-esteem, a healthy body weight, and achieving happiness and satisfaction with life. And even if we have been doing things all wrong our entire lives, if we change today, we still stand a better chance tomorrow and for the rest of our days to reverse a significant amount of the damage we may have done.

Since a person's lifestyle is so important, it makes sense to know not only the general features associated with a healthy lifestyle, but specific activities that can be incorporated into one's daily routine to help achieve it. In this chapter, I will focus on discussing a variety of what I believe are the most important of these. While it may be tempting to rank these in terms of order of importance, we can avoid playing that game by simply stating that each of these (diet, exercise, sleep, and stress management) play a key role as part of a healthy lifestyle.

PURPOSEFUL NUTRITION

In Buddhist practice, one learns to be contemplative ("mindful") regarding the food that is put into the body. This is in part because food choices impact our health and well-being, both positively and sometimes negatively. This is essentially the idea behind **purposeful nutrition**. It deals with thinking about the consequences of what we eat and drink, such that we should have a purpose or goal in mind connected to this behavior. In other words, if our goal is metabolic health, then we need to eat in such a way that we will achieve it rather than to jeopardize it. And the same holds true if our goal is to build strength and muscle, combat aging, and improve our ADLs and overall functioning.

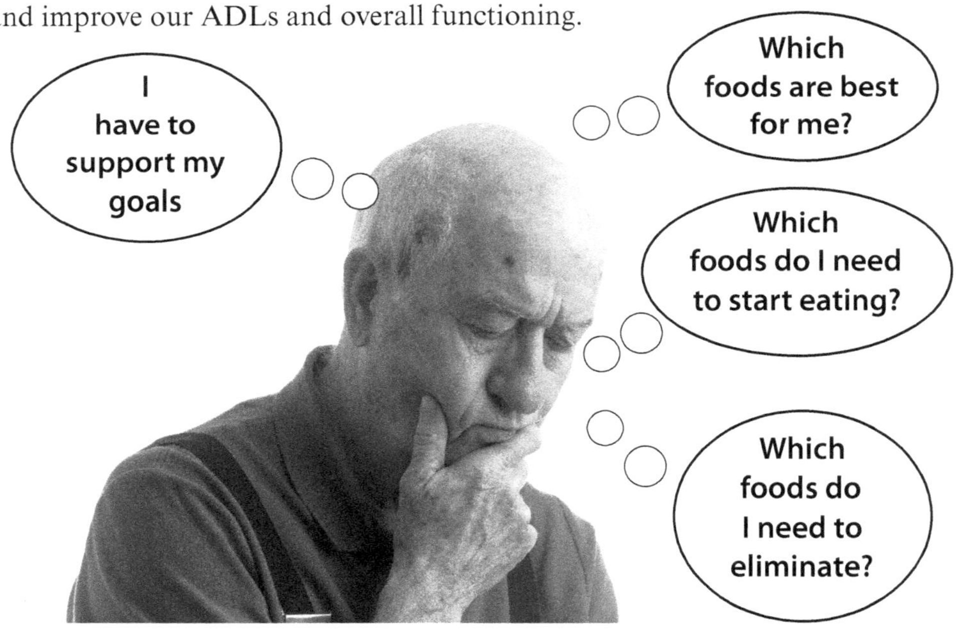

▲ **Figure 5.2** Putting your purposeful nutrition plan into practice is aided by establishing your goals and thinking mindfully – critically – about the foods that you will be routinely consuming from now on. Certain "old favorites" may need to be minimized or eliminated if they are contrary to your goals, and your food environment at home will probably require an overhaul.

Purposeful eating implies that there needs to be a reason for the way we eat as older men – which in turn impacts the specific foods (and beverages) and the amounts consumed. Otherwise we will simply leave things to chance and run the risk of eating foods that taste good or are comfort foods, but might also be junk foods, foods high in fat, sugar, sodium and so forth – to the detriment of our aging bodies. Just as we needed specific goals with respect to strength training, we will be discussing goal setting with respect to nutrition. Can you see at the outset that having as a goal to be more thoughtful regarding food choices will support our healthy weight, muscle building, anti-aging, and increased functionality outcomes?

The latest nutrition research is meant to be reflected in the Federal dietary guidance (USDA's MyPlate tool and the U.S. Dietary Guidelines for Americans). However, not everyone understands or makes the effort to investigate how to make use of these recommendations. Instead, many people resort to buying dieting books, listening to talk show celebrity guests, late night infomercials, or following the latest fad diet that their friends are trying. The strategies may not be the best when it comes to making sound lifestyle choices.

What is a Good Diet for Older Men?

What is the right eating plan (diet) for us to follow? This is not a simple question to answer, because it depends on a variety of factors, including: (1) your current age, weight, and BMI; (2) your current percent body fat; (3) how much weight and fat you need to lose; (4) your current muscle mass and how much you want to gain; and (5) what your training (and overall health and fitness) goals are.

In general, many of us will simply want to lose some weight and body fat, and gain some muscle and strength. We want to look good with our shirt on or off. We want to improve, rather than continue to decline, in terms of physique and strength. These are straightforward and achievable goals, and such improvements will boost our confidence and mental outlook, making life more enjoyable.

Back to our eating plan. There are a zillion diets and diet gurus out there with their own pet theories and advice. I would say that while many may be sincere and believe that they are offering good advice, it is easier for me to believe such is the case when the person is not peddling a particular supplement or plan that you must purchase only from them. In such cases, there may be a scam in progress.

What we can initially appreciate is that we are all individuals with slight to large differences in our physical and metabolic systems, and overall health. Everything from the amount of body fat and muscle to blood glucose and triglyceride levels displays uniqueness. However, a sound diet can transcend such differences and be broadly applied to almost anyone. I can suggest what a good eating plan for older men should be like, and it is as follows:

1. favors the anabolic metabolic state rather than the catabolic one in terms of protein turnover
2. is supported by adequate water intake
3. is adequate in total calories to achieve and maintain the proper weight for height
4. provides a reasonable allocation of calories coming from protein, fat, and carbohydrate – and includes the extra protein needed by older men
5. provides all the essential vitamins and minerals, all the essential amino acids, and all the essential fatty acids that are required for optimal health

6. is low in heavily **processed foods**
7. includes a substantial amount of plant-based foods - a sensible variety of fruits, vegetables, and grains in order to obtain phytochemicals (natural molecules that are not recognized as nutrients but are demonstrated to provide benefits to human health)
8. contains a few supplements that have solid research behind them
9. is low in or does not contain "junk foods", fried foods, and "fast foods" high in any or all of the following: total/saturated fat, simple sugars, and salt/sodium.
10. is low in or does not contain alcohol and sweetened beverages (such as soda pop, sweetened juices, ice cream sodas, malts and milk shakes), nor does it contain artificially sweetened beverages

THE OLD MAN MUSCLE EATING PLAN

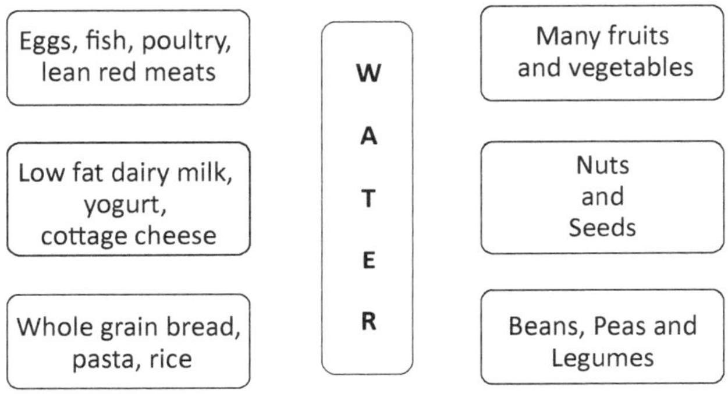

▲ **Figure 5.3** The Old Man Muscle Eating Plan includes water plus a balance of foods that will provide the older man with the appropriate nutrients in sufficient amounts. When followed, it will serve as an anabolic diet that promotes muscle growth, as well as being an anti-inflammatory and anti-aging diet. Careful selection of specific food items and amounts will ensure balance and the ability to achieve and maintain metabolic health and the proper body weight. While all of these foods are important in a balanced diet, if we can emphasize water intake and fill up with the foods on the right side of the figure more than we typically do, we will find it difficult to overeat the foods on the left hand side. Maintaining a healthy weight will be easier – and we will not have to count calories.

I have thought long and hard about how to answer this question – and really, what to emphasize throughout this chapter. My doctoral training as a nutritionist, my experience as a food science and obesity researcher and author, and also my background working for the USDA Federal nutrition assistance programs, would have me focus on the following: list and explain all the macronutrients, micronutrients and their functions; discuss the energy balance equation and how to lose weight and maintain a desirable body weight based upon calories in vs. calories out; discuss the USDA's MyPlate tool and the U.S. Dietary Guidelines for Americans, and how to use them.

USDA MyPlate Tool

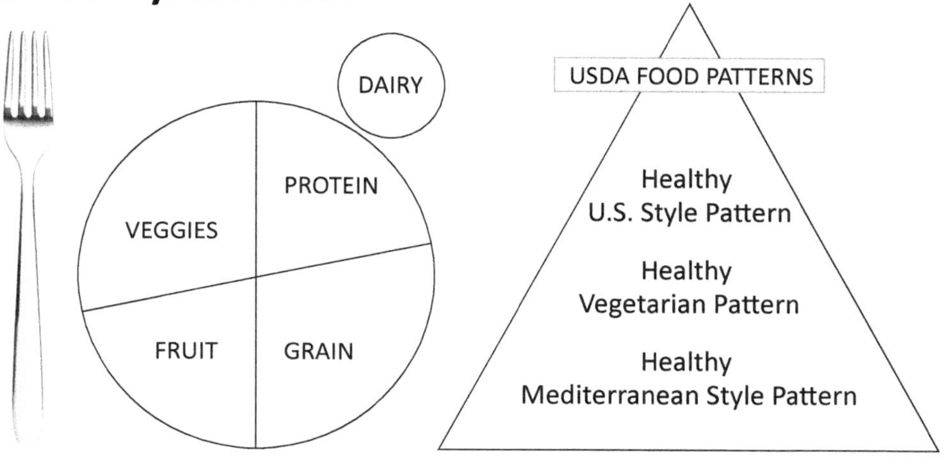

▲ **Figure 5.4** The USDA MyPlate and Food Patterns are educational tools meant to help Americans incorporate the U.S. Dietary Guidelines into their daily diet. You notice that protein – a nutrient - is one of five categories, the other four being food groups. This is because consuming protein from a variety of foods is important. Foods made from meat, poultry, fish, beans, peas, eggs, nuts and seeds are considered part of the protein group. The three food pattern options identify daily amounts of foods, in nutrient-dense forms, to eat from five major food groups and their subgroups. You can read details of the MyPlate and Food Patters at the USDA.gov website.

While the topics just mentioned all appear in most standard nutrition textbooks including my own, I have decided to be a bit more creative and provide information targeted to older men. This, after all, is the whole point of the Old Man Muscle handbook. We face specific challenges and have special needs. I therefore have not written a nutrition section full of dietary generalities – instead, we will examine the relevant nitty gritty questions.

What are our dietary (nutritional) goals?

Just as in strength training we have goals, so it is also with respect to our eating. We have already touched on this by saying that we want to eat in such a way that we (a) attain sound metabolic health, (b) achieve a healthy weight for our height, (c) support our strength training and muscle building, (d) slow down the aging process, and (e) remain functional into our advanced age. These parameters represent outcomes that should result if we work hard to achieve them. Let's remember that anything worth having is worth working hard for. Look at the list from (a) to (e) again – do you really want these things? If so, work hard at achieving good nutrition – and they will be within your grasp.

Clearly, eating for pleasure versus eating with the goal in mind of becoming stronger are two different things. We will ultimately commit to one or the other – we cannot do both and expect to change things around. Mindful eating will get us there – but mindless eating will not! We are looking for a diet that is healthy and clean, is anti-aging, and is effective for older men to enable a muscle and strength transformation to take place.

Setting goals is potentially one of the most beneficial activities that we can engage in as we age. Of course, it is important to follow through and meet them as best we can once we've identified them. We should understand the difference between establishing **outcome goals** versus **behavior goals**. The outcome is what you want to have happen, while the behavior is what will get you there.

BEHAVIOR GOALS	OUTCOME GOALS
KEY FEATURE: We focus on the process, activities, behaviors that enable us to achieve our goals.	**KEY FEATURE: We focus on the results.**
"I will increase my protein and plant food intake, while eliminating fast foods and sweetened drinks. "I will strength train three times per week."	"I will gain enough muscle mass that my buddies notice and make comments." "I will lose 20 pounds after one year of training." "I will look good in shorts and a T-shirt."

▲ **Table 5.1** Examples comparing behavior and outcome goals, and how they are related.

Establishing the right food environment at home

▲ **Figure 5.5** Since adults still consume a majority of their daily energy intake at home, it is a key place to establish healthy eating behaviors. The quality and quantity of the foods available thus play a significant role in eating behaviors and health. Access to the right foods, which need to be consumed keeping proper portion sizes in mind, becomes very important. Generally speaking, meats and dairy products need to be consumed in moderation compared to fruits and vegetables, which for the most part we can eat as much of as we like.

The home food environment, coupled to the food behaviors that we engage in when we eat out, will have a significant impact on meeting our strength and overall healthy aging goals. This is because of the close association between diet and these two key parameters. For example, if we consume a wide variety of plant-based foods in addition to animal products that provide us with high quality protein, this will support our strength training and muscle protein synthesis goals, as well as being anti-aging. If we cut down on refined carbohydrates and fat from fried foods and desserts, in favor of complex carbs (whole grain and fiber) and fats from nuts, seeds, and dairy, these will be more metabolically healthy for us. Since the availability of specific foods at home (the home **food environment**) will determine to a large measure our dietary intake, and is under our control, it becomes our responsibility to populate our refrigerators, freezers, shelves and pantry with the right foods.

How should I divide my macros?

We need to be fundamentally cognizant of the fact that the foods we eat provide one or typically two or three of the **macronutrients** the body needs in various proportions. The three macronutrients are protein, fat and carbohydrate. Since these constitute the largest amount of what we eat, they are considered "macro" nutrients. Macronutrients provide energy to sustain life so we can move and function, and contribute to body tissues and organs like brain, lung, liver, and muscle as well as composing regulatory substances like enzymes and hormones. It has become fashionable for people to use the term "macro" when discussing the impact of the macronutrients upon weight gain, weight loss, and health.

On the other hand, the body also requires **micronutrients** – various vitamins (think of vitamins C and E) and minerals (like calcium and iron). These assist in obtaining energy from the macronutrients and may be involved in enzymatic and hormonal processes, so they too are supportive of proper functioning.

In addition to the macronutrients and micronutrients just presented, many plant-based foods contain naturally occurring molecules that are not proteins, carbohydrates, fats, vitamins, or minerals but have been shown through scientific research studies to be of potential benefit to health in specific ways. These kinds of substances are termed **phytochemicals** ("phyto" meaning plant). For example, there are numerous types of polyphenolic antioxidant compounds found naturally in fruits, vegetables, berries, beans and legumes. Products such as coffee, tea, and chocolate also contain these substances.

Certain animal-based products, while not containing phytochemicals, may contain additional health-promoting substances that are neither macronutrients nor micronutrients. Fermented dairy products and fermented meats, for instance, contain specific acids and other bioactive substances that may function in this regard. Also, let me make special mention of red meat. While I am not advising that we eat a lot of red meat all the time due to the fact that we can obtain protein from many other foods, beef brisket deserves our attention. I say this for a few reasons: a 3-ounce (quarter pound) serving of brisket is packed with 28 grams of quality protein. But many folks are unaware that brisket also contains a healthy kind of fat called **oleic acid** (a monounsaturated fatty acid) that increases HDL (good cholesterol), and lowers LDL (bad cholesterol). This reduces our risk of heart disease. But don't go overboard with brisket or any one food for that matter. Variety is the key, and we have so many good protein choices in addition to brisket to include in our diet so that we get better coverage of all the nutrients, not just protein and one type of fatty acid – important though those are.

The manner in which we can obtain nutrients and bioactive substances is typically by eating raw (uncooked, unprocessed) foods, minimally processed foods, and also processed foods. The two major sources of food for us on this planet are plant-based foods and animal-based foods. The conventional wisdom for optimal health is to emphasize plant-based foods (fruits, vegetables, grains, nuts, seeds, legumes, etc.) and less so the animal-based ones (dairy, eggs, meats, poultry, seafood, etc.). It is important to include raw fruits and vegetables and minimally processed plant-based foods (and juices) as long as they are safe to consume (not contaminated with pathogenic microorganisms that can cause disease).

Typically in the animal-based category, we should never consume raw meat, raw milk and cheese, or raw eggs because of the risk of foodborne illness due to contamination with organisms like E. coli and Salmonella. Processing in terms of added antimicrobials, thermal (heat) treatment such as pasteurization, refrigeration and freezing makes a diversity of animal products like milk, deli meats, ground beef, salmon, and eggs safe and nutritious to consume.

The proportion of each macronutrient that should be consumed is the primary point of discussion, and opinions vary. To start thinking about this, we can focus on a few specific foods, and then expand our thinking to consider a complete day of eating. Let's take a slice of bread, a can of tuna fish, and a banana as examples. I can easily search the internet for the nutrition information for each of these, and when I do, I am able to report the following:

Calories	Protein	Fat	Carbohydrate
One slice of whole wheat bread (38 grams, approximately 90 calories)	provides 4 grams of protein	1 gram of fat	16 grams of carbohydrate
One 5 oz. can of tuna (113 grams, approximately 100 calories)	provides 24 grams of protein	1 gram of fat	0 grams of carbohydrate
One medium banana (118 grams, approximately 105 calories)	provides 1 gram of protein	<1 gram of fat	25 grams of carbohydrate

So, what we see is several things. First, each of these foods differs with respect to their macronutrient composition. The tuna fish is by far the best source of protein, while the banana is mostly carbohydrate. Being a raw, unprocessed fruit, bananas contain healthy carbohydrates. The whole wheat bread is somewhere in between. What else - we could calculate the percent of calories coming from each macronutrient for each of the foods if we take into account their respective energy (calorie, or "physiological fuel") values: 1 gram of protein = 4 calories; 1 gram of carbohydrate = 4 calories; 1 gram of fat = 9 calories.

For example, let's calculate the calorie split for the bread. The protein component is 4 grams x 4 calories/gram = 16 out of the total 90 calories, which is about 18%. The carbohydrate component is 16 grams x 4 calories/gram = 64 calories, which is about 71% of the total. The fat provides 1 gram x 9 calories/gram =10% of the total calories. So the macro split for this one food is 18% protein, 10% fat, and 71% carbohydrate.

The point is, researchers have for years been investigating not just the nutrient analysis of individual foods, but what the overall daily diet should be in terms of macronutrient proportions, or "split" based upon the total number of calories coming from all foods consumed on a daily basis for a person having a normal healthy weight for height. The USDA guidelines and the Old Man Muscle suggestions on this are provided in the figure below. An interesting question would be, "is there any metabolic advantage to variations in macro split that perhaps are high in protein, or low in carbohydrate, or high in fat, or low in fat?" It is a complex question, because while one diet might provide an advantage in terms of cardiovascular health, it might not be the best for blood sugar control or building muscle, or as a weight loss or anti-aging strategy. The USDA percentages are for overall health of adults, while the Old Man Muscle percentages are more focused on older men, hence the protein emphasis. It should be possible to tweak either plan based upon the consumption of foods specific for the various potential purposes (anabolic, anti-aging, anti-inflammation), over and above simply calorie contribution for body weight maintenance.

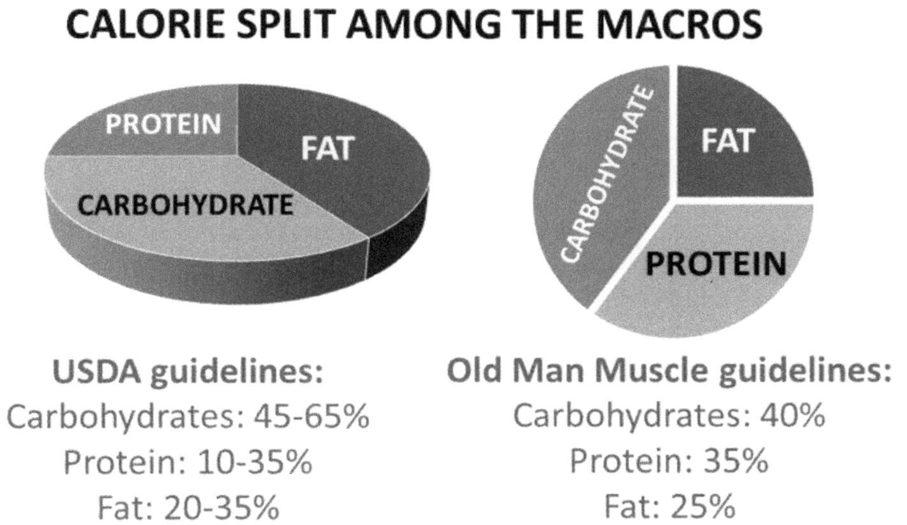

▲ **Figure 5.6** Suggested Old Man Muscle calorie splits (right) among the macronutrients are within the range of USDA values (left) for protein and fat while being a bit less for carbohydrate. Protein requirements increase with old age due to anabolic resistance; fat intake should emphasize healthy fats from eggs, meats such as brisket and fish oils, and nuts and seeds to help boost testosterone levels and the immune system. Complex carbohydrates primarily from plant foods (fruits, vegetables, cereals, grains, legumes) make up the largest percentage of calories each day to provide fiber and associated phytochemicals.

Before leaving the topic of macronutrients, I want to suggest something that may or may not be obvious. If we adopt a high protein diet, eat several meals per day (minimum of three but for larger men, up to five or six to get enough protein), emphasize a low processed food diet with plenty of plant-based foods to accompany our animal proteins, and drink an adequate amount of water, we will not have to count calories or macros because we will have eaten healthfully just through the proper food choices. Life is complicated enough – why not keep things as simple as possible? This approach can work for a person who is at a healthy weight and seeks to maintain it. For a person who needs to lose weight, it may be possible to lose weight (and body fat) more rapidly on an intermittent fast, which however has some challenges associated with it if you are trying to build muscle at the same time as losing weight and eating in the manner I just outlined. I will discuss intermittent fasting in a moment – please continue reading.

Is a low-carb diet for me?

Carbohydrate (carbs) is recommended to make up the greatest percentage of the macros in your daily intake. There has been interest of late in the **low carb diet**, as a weight loss strategy. The bottom line with carbs is as long as the emphasis is on eating complex carbohydrates like fiber and whole food sources (fruits, vegetables, peas, beans, legumes, whole grain breads, cereals, nuts, seeds, rice and pasta) rather than simple sugars found in processed high-carb foods (cakes, donuts, pies, pastries, candy, sugary beverages, and other sweets), you can follow a healthy low carb diet. There are powdered supplement products on the market (example: various brands of "Super Reds" and "Super Greens") that can be added to beverages by the spoonful to provide important phytochemicals from fruits and vegetables missing from typical low carb diets that you may wish to supplement with. But I do not recommend a low carb diet as a permanent lifestyle, mainly because of the fiber and nutrients they provide. Added to this is the fact that we need to increase our intake of plant-based foods (which coincidentally are high in carbs) as an anti-aging strategy. These foods contain natural bioactive molecules that are anti-inflammatory and anti-oxidant. Refined carbs do not have them.

There are some contraindications for a low carb diet. There is a risk of hypoglycemia, so anyone with diabetes should be cautious. Also realize that when we work out, we use up some stored muscle glycogen (the source of the important carb metabolically speaking: glucose). This will cause our muscles to shrink in size. The best way to replenish it as an aid to muscle building is to have sufficient carbs as part of the diet. You can tell if you are good with your muscle glycogen stores by how your muscles look. Check your flexed biceps; are they rounded and full, or flat? Without the glycogen (and water that is attracted into muscle tissue with glycogen), it will be flat city, my friend.

A final word to my fellow older men on this. At this point in our lives, we need to change how we eat carbs - away from cakes, cookies, donuts, pies, sweets, and so forth and toward fruits, vegetables, legumes, nuts/seeds, whole grains (including breads, rice and pasta). Consuming these foods is the better way to replenish glycogen, and not with sugary drinks, candies, and cookies! I know it is hard to do, but so is drinking enough water, and strength training! If you commit to doing one, commit to doing all. Otherwise, don't expect things to change.

Is the "keto diet" for me?

The ketogenic or "keto" diet has become very popular recently. Before discussing it, I wish to mention that our diet is obviously not the only factor contributing to weight loss.

Physical activity, stress reduction, adequate sleep, socializing and mindful eating all are part of the equation. It is important to address each, and in terms of a diet to follow, there are the ones that you do short-term for "quick" fat and body weight reduction, followed by the long-term lifestyle diet that you will follow permanently, for maintenance purposes. The ketogenic (or "keto") diet, along with most others, can help jump-start weight loss, but then you should shift over to a steady, healthy weight diet that you can follow the rest of your life.

The ketogenic diet is a low carbohydrate ("low carb") higher fat and protein diet that restricts things like breads, cereals, rice, pasta, cereals, and bakery items in favor of eggs, meats, seafood, nuts, and unsweet dairy. When this diet is properly applied, you are to consume no more than 30 grams of carbohydrate (simple or complex) each and every day, a very restricted amount, as many men typically consume 200+ grams of carbs daily. The 30 grams of carbs should come primarily from non-starchy fruits and vegetables with good fiber content. Also, there is no room for sweetened beverages, including 100% fruit juices.

The reason to be so restrictive of carbohydrates is in order to use up or deplete the body's storage of glycogen. Glycogen is a storage polymer of glucose, the preferred fuel of body cells such as brain (which needs about 100 grams daily). Since the brain will not have access to much glucose on the keto diet, it will be forced to use the only other fuel source it can metabolize for energy: **ketones** (or "ketone bodies"). This in fact gives the diet its name. Where do ketones come from? When we consume very little carbohydrate, first liver and then muscle glycogen stores become used up. Since body cells other than the brain can use fat, energy metabolism shifts away from using stored glycogen and ingested carbs (there will not be much) to burning body fat for energy. When fat is oxidized in this way, it generates ketones, which is a very good thing because brain cells (and also muscle) can metabolize ketones for energy needs.

It takes up to one week for the person using the keto diet to use up available glycogen stores. At that point, with little glucose around, insulin levels drop while glucagon levels rise. This is a hormonal reversal from what is typical, and improves insulin sensitivity. Also, and of significance, the body begins to deplete stored fat, generating ketone bodies and putting the individual into a state of **ketosis**. Weight loss can be rapid, and is mostly water weight at the start. It is important to remain fully hydrated by regularly drinking water. Due to a process called gluconeogenesis, the body can synthesize glucose from protein and fat fragments, and these can continue to fuel muscle. This is what generally happens at the start of the keto diet. However, a person needs to remain more in a state of ketosis than **gluconeogenesis**, in order to burn excess body fat. This is challenging to achieve when you follow a high protein diet such as we do as older men in order to gain an anabolic edge and fight sarcopenia. The duration of time to remain on the keto diet will vary person to person depending upon various factors, but it is typically safe for a month or more. I would not advise an older man to remain in the ketogenic state for longer than two months. One can always try it out for a month and apply this diet more than one time during a calendar year to help achieve weight loss goals. Since the long-term effects of this diet are not yet known, caution is advised.

On this diet, which in reality is difficult to follow because it is so restrictive of carbs, when your breath begins to smell of acetone, then you know you are becoming "ketoadapted" and in a state of ketosis. You have shifted from glucose burning in body cells to fat and ketone burning. If you choose to try this diet, realize that it is not advisable to do so if you are a diabetic, unless supervised by your physician. The risk of hypoglycemia is a real concern, so the doctor will monitor your blood glucose levels. Also, you would want to ease gradually down on the carbs, by about 10 grams per week in order to prevent hypoglycemia and what has been referred to as "the keto flu". Shifting immediately from a medium or high carb diet to the keto diet is not smart. Eating frequent, small meals is also suggested, never going more than 3 or 4 hours while awake without eating.

Is intermittent fasting for me?

You may have heard about **intermittent fasting** – but what exactly is it? Some have called it "intermittent energy restriction" or "time restricted eating", which provide clues to its meaning. Intermittent fasting is an approach to subjectively and consciously control what scientists term "feasting versus fasting", and has to do with the timing of our eating periods. Feasting means the hours in the day when we consume our food, including all meals and snacks (total food and beverage intake). Fasting has to do with the hours in the day when we put absolutely no food or drink – other than water or non-caloric beverages (e.g. unsweet tea) into the body. In research on animals, studies use the term **TRF (time restricted feeding)**. This is how the connection between intermittent fasting and improved metabolic health was first learned.

Is Your Lifestyle
Helping or Hindering Your Body Cells

▲ **Figure 5.7** Our body cells are fundamentally important - and only we can take care of their needs. They enable us to perform all of life's activities that we enjoy. From our brain cells to our muscle cells, they carry out all the intricate details of metabolism so we not only stay alive, but thrive. What our cells ask in return is: enough food, water, and rest so that they can do their jobs to the fullest for as long as possible. A couple of glasses of water per day isn't enough; a fast food diet will fail to deliver the necessary nutrients; and if we eat with abandon on and off all day long our cells are constantly having to work to pump out enzymes, hormones, and chemical signals to keep up, with no down time for recovery and rejuvenation. Intermittent fasting may help if you choose to implement it from time to time in your schedule.

The idea of fasting has been around forever. In some religions, a fast is used as a way to cleanse oneself, shifting the focus away from the needs of the physical body toward the mind, soul, and **spiritual** dimension. Scientifically speaking, fasting allows all the cells of the body to rest and recover from their constant 365-day per year activities. Cells make up tissues, and tissues comprise the organs of the body – and the key systems (brain and CNS, circulatory, digestive, endocrine, neuromuscular, etc.). It is thought that fasting "down time" allows for rejuvenation and resetting of the optimal functioning rhythm to occur in these body systems. But just how much of this down time is required in order to realize benefits?

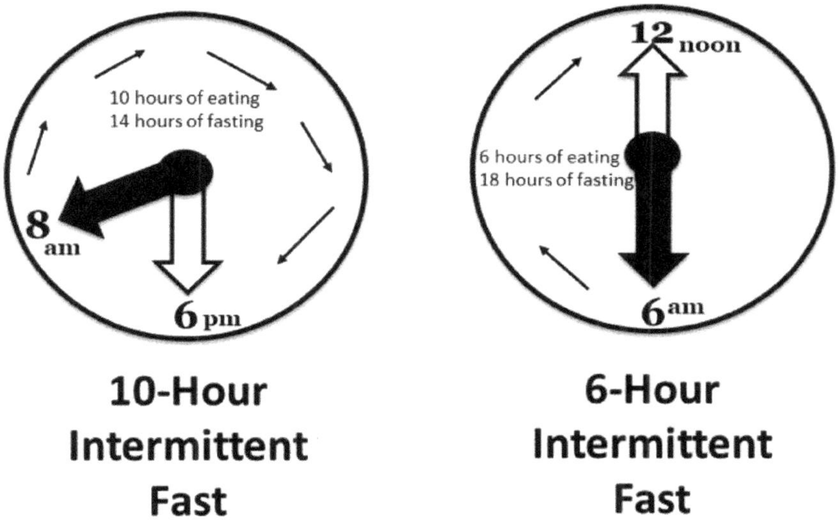

▲ **Figure 5.8** Intermittent fasting. A 10-hour plan entails having your first meal of the day at 8 a.m., for example, and making your last meal of the day by 6 p.m, which is ten hours later. This means that you eat within a 10-hour window, and fast the remaining 14 hours. Research has shown promise for weight loss with this strategy. Further metabolic benefits are associated with extending the fast, which however narrows the eating window still further. The quite restrictive example shown to the right is a 6-hour eating window. For example, you could take your first meal of the day at 6 am, and by 12 noon you complete your last meal of the day. You then fast for 18 hours.

Intermittent fasting is distinguished from regular fasting, which is total avoidance of food for one or more days. With an intermittent fast, the period of the fast is a much more brief time interval than in a regular or total fast. Total fasting leaves one only with water intake over a span of days or more. It is potentially damaging to the body the longer the duration, upsets the natural circadian rhythm that body cells follow and function best with (to secrete hormones, carry on metabolic activities necessary for health and life, etc.), and should only be practiced with caution. Intermittent fasting is much safer, although at the end of this discussion I will mention a few concerns.

Usually, people tend to eat within a twelve hour window throughout the day, while fasting several hours before bed and throughout the night during sleep, to account for the other twelve hours in a day. Intermittent fasting practices vary, but they typically alternate shorter periods of eating with longer periods of fasting in a 24-hour period. One approach restricts eating to 10, 8, or even 6 hours per day. The more restrictive the eating phase, the more effective the plan regarding weight loss and metabolic health improvement. Another approach is a bit different and allows a sensible diet during the week, while severely restricting food on the weekends. The reason that intermittent fasting works, beyond allowing for cellular rejuvenation, is that it apparently triggers a metabolic switch that shifts energy nutrient dependence away from glucose in favor of fats and ketones. This change, over time, helps one burn fat, lose weight, and improve glucose regulation. These are features in common with the keto diet. There also can be an increased resistance to oxidative stress, a decrease in blood pressure, and diminished carbohydrate food cravings as well.

However, there is no doubt that the idea of fasting, in our food indulgent and pleasure dominated culture, can be quite challenging to actually put into practice as a lifestyle change. Even a small amount of added cream in coffee or sips of a sweetened beverage during the fasting period is enough to break it, nullifying the benefits. Adherence to a plan like intermittent fasting can be very difficult or even impossible for some due to their emotional connection to food and eating (people with eating disorders, especially).

While intermittent fasting is one strategy that can be tried by many in order to lose weight and become more metabolically healthy, it is not for everyone. If you are underweight, diabetic, have an eating disorder, or are in your 80s or 90s, for instance, it is not recommended. The reasons include the risk of hypoglycemia (low blood sugar) due to the restricted carbohydrate calorie intake, which may predispose one to light-headedness, black-outs, and falls. I would also add that similar to the ketogenic diet, one may experience low energy when working out, and struggle to perform up to par due to low glycogen levels. In addition, maintaining restrictive eating windows for long periods of time may actually do harm to the cardiovascular system. Additional possible adverse effects in older adults, potentially including dehydration, heart arrhythmias, and gallstones - really need to be more completely studied. As renowned researcher Dr. **Len Kravitz** has advised, all the facts are not yet in regarding how best to apply intermittent fasting ("no one really knows which type of fasting is best for different goals – whether that's fat loss, muscle preservation, disease prevention, or longevity"). Trial and error using self-experimentation, may be the best way to go, starting with the least difficult 10/14 split of feeding/fasting, and becoming more restrictive over time.

Can I train and eat to simultaneously gain muscle and lose fat?

One key issue I want to quickly address now is, can a person train and eat in such a way as to gain muscle and simultaneously lose fat? The reason this question is important is that when the majority of people go on diets to lose weight (e.g. with a standard caloric deficit of 500 calories per day) they unfortunately lose muscle mass right along with fat mass and body weight. This is counterproductive to our goals as older men. Research has been conducted to examine this issue. What has emerged is the following: weight (strength) training is essential to maintaining muscle mass while dieting; dieting without strength training leads to muscle loss; being in a caloric deficit is essential to losing fat; extra dietary protein during dieting aids in losing fat; 2 grams of protein per pound of body weight is superior to 1 gram/pound in terms of adding muscle; protein overfeeding is unlikely to cause any gains in body fat, and appears to actually reduce body fat, and does not harm the kidneys or other body organs. (unless there is underlying disease present).

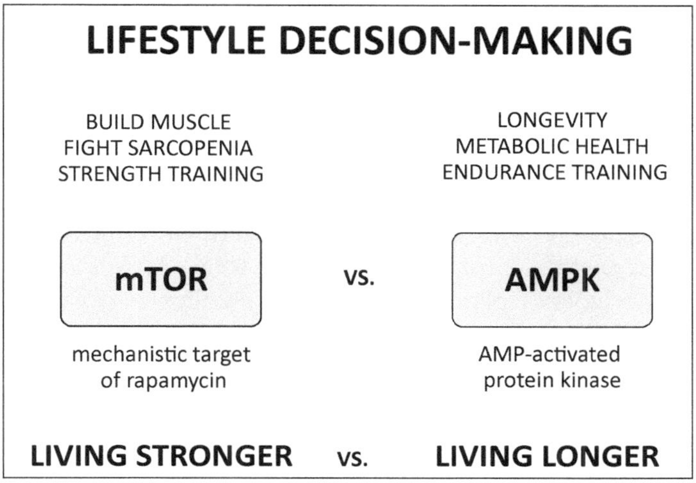

▲ **Figure 5.8** We have the challenge as older men of striking a balance between a lifestyle that favors building strength and muscle in order to fight sarcopenia vs. a lifestyle that favors longevity and is anti-aging. Building muscle is associated with the anabolic mTOR pathway while extending lifespan and improving metabolic health favors the more catabolic AMPK pathway. Both mTOR and AMPK are regulatory enzymes that play key physiological roles. Functionally they are antagonists since AMPK can inhibit the action of mTOR. The way that we train and the way that we eat can favor one over the other. Is it possible to strike a balance?

Related to this issue is understanding the basic roles of two key regulatory proteins (enzymes) that work in opposition to one another: mTOR and AMPK, and the metabolic and cell signaling pathways they are involved in. Much research has determined that mTOR is the master regulator of cell growth, and is influenced by the presence of nutrients and also exercise factors. If our goal is to increase muscle mass as we age by enhancing MPS, we need to stimulate mTOR, which unfortunately becomes more difficult due to a blunted response as we age. We overcome this by following an anabolic strength training program along with an anabolic diet. Such a diet would provide a slight daily surplus of calories, and tend to favor a high protein intake with plenty of BCAAs, relatively high fat intake, and only a moderate amount of carbs. This could be approximated with a fairly even 35:35:30 macro split, for example (protein:fat:carbs). Or, use the Old Man Muscle macro split (35:25:40) which allows for more metabolically healthy plant-based foods as carbs. But we don't want to chronically overstimulate mTOR, as this may promote the growth of cancerous cells and also lead to premature aging.

Meanwhile, AMPK can be stimulated by employing fasting strategies, (recall intermittent fasting), and an overall lower calorie diet. This approach increases fatty acid uptake and oxidation to help you lose fat, while decreasing the formation of new fat cells (adipocyte lipogenesis) and increasing fat breakdown (lipolysis). The anti-aging aspects that are suggested to kick in are reduced oxidative stress, reduced chronic inflammation, improved blood pressure, and improved glucose regulation (all indices related to improved metabolic health). However, as with mTOR, due to aging we experience a blunted AMPK response. The strategy to work around this blunting is essentially the opposite lifestyle in terms of diet and exercise to the mTOR approach. We would emphasize cardio workouts, and a calorie-deficit diet that is lower in protein and higher in carbs - something like a 15:25:60 macro split between protein, fat, and carbohydrate. The key here may be more related to the magnitude of caloric deficit rather than the macro split, because a hypocaloric diet will provide less glucose and fewer amino acids, which results in less insulin and less MPS, and is linked to longevity. But I think the reader can see that this is an anti-anabolic (in other words, catabolic) lifestyle, and is therefore the opposite to what the Old Man Muscle lifestyle is all about.

What can we do if we like the idea of a longer, healthier life but also want muscle functioning into advanced age? We can follow an anabolic training and dietary pattern, focusing on a clean diet that de-emphasizes processed foods while focusing on healthy animal protein foods and lots of plant-based foods. Consider an occasional intermittent fast and the possible addition of some anabolic, antioxidant and anti-inflammatory supplements. With this approach, it may be possible to simultaneously achieve both anti-aging and muscle and strength building outcomes. We can be stronger and live longer!

Should I avoid processed foods?

Processed foods – are they all bad? Definitely not, especially when you consider that milk is a processed food, eggs are a processed food, yogurt is a processed food, and so are frozen foods that may be nothing more than fresh fruits and vegetables that have been packaged and flash frozen to preserve their nutrient content. Then, too, there are the "healthy" and "high protein" frozen bowls and meals that have recently hit the market – not to mention orange juice and the high protein, high fiber breakfast cereals. If we were to focus on consuming such foods instead of greasy fast foods and sugary beverages – how our muscles would love us, and how nicely our clothes would fit!

In my view, there are three general kinds of processed foods. They are: heavily ("ultra") processed foods that contain many ingredients or additives listed on the food label; moderately processed foods having fewer chemical additives; and minimally processed foods (e.g. sliced apples treated with citric acid and sealed in "grab and go" bags). While processing has definite uses in the food industry that are important such as extending product shelf life, making foods safe to consume, and preserving vitamin and nutrient content, palatability to consumers often means that the processor adds more sugar, more salt, and more fat. Too much of these "limited extras" can make a food in any of the processed categories less healthy. We should also keep in mind that research has demonstrated that people whose diets emphasize heavily processed foods are associated with increased calorie intake and weight gain. From this, we should learn to choose a diet that is relatively "clean" - low in processed foods, while emphasizing fruits, vegetables, and plant-based foods in addition to quality **protein sources**.

Should I avoid eggs and dairy?

No. Scientific evidence is mounting in favor of consuming dairy products as well as eggs in order to maintain good metabolic health. As you know, these are animal products and contain cholesterol and saturated fat, as well as high quality protein. For years, the American Heart Association and others have been warning us not to eat a lot of eggs because of the cholesterol in the yolks. The food industry responded by making available "egg beaters" that function as a substitute. But with the latest research in, the warnings regarding cholesterol in foods have mostly been silenced. It is now okay to consume an egg a day, and potentially more. An egg per day is said to pose no increased risk for heart attack, stroke, or any other type of cardiovascular disease. Of course it makes sense to have your cholesterol levels checked anyway, on a regular basis, to be sure that your blood levels are in the normal range for not only cholesterol and triglycerides, but for other health and disease markers.

Egg yolks – which have been the subject of much abuse – actually are rich in polyphenolic compounds (lutein and zeaxanthin) that make them anti-inflammatory foods. The situation here is more complex than with cholesterol, however, because some egg bioactive components are pro-inflammatory while others are anti-inflammatory. Importantly, eggs are loaded with protein and nutrients including omega-3 fatty acids, as well as growth factors. So on balance, due to the nutritious aspects of the protein in egg, the omega-3 fatty acids, the anti-inflammatory molecules they possess, and Vitamin D, eggs are regarded by scientists as wholesome and inexpensive functional foods that have far more in their favor than against them. Relevant to our strength training, studies have shown that ingestion of (cooked) whole eggs after resistance exercise resulted in greater muscle protein synthesis than the ingestion of egg whites alone.

Regarding dairy foods, research has demonstrated that consumption of dairy products supports healthy weight and lean body mass, and decreases one's risk of hypertension and potentially also type 2 diabetes. Milk for instance possesses bioactive peptides that can reduce blood pressure and enhance endothelial status (the lining of blood vessels) by inhibiting ACE, which stands for **angiotensin-converting enzyme**.

As we have learned, in order for a person to be in good metabolic health, he should not be diagnosed with chronic disease symptoms via physical exam and/or blood analysis. What is termed **metabolic syndrome** results when a cluster of risk factors (minimum of three) are observed simultaneously in a person. These include increased waist circumference (large gut), elevated blood lipids (triglycerides), reduced HDL-cholesterol (the "good cholesterol"), high blood pressure (hypertension), and elevated fasting blood glucose (also called insulin resistance).

We have mentioned inflammaging as a chronic low-grade inflammatory state that accompanies (and likely causes to an extent) aging. Alzheimer's disease, cardiovascular disease, Type 2 diabetes, obesity and cancers have been associated with inflammaging. Minimizing inflammation, therefore, is one key to healthy aging. Newer research indicates that certain dairy foods, including whey protein and yogurt, might help fight inflammaging. We need to remember that whey protein is naturally present in milk and milk products (but not in cheese since the whey is drained out). Regarding our strength training, whey protein after exercise helps us to build muscle and maintain a healthy bodyweight, two criteria critical to healthy aging.

WATER INTAKE

Someone recently told me that strength training and proper water intake go together like a needle and thread. This is because when we fail to be properly hydrated, there is impairment of muscle protein synthesis and ultimately, function. This includes decreased exercise performance, decreased number of repetitions, increased perceived exertion, and hindered recovery. It's also the case that every time we eat, we should take in water. The very fundamental processes of digestion, absorption, transport, utilization, and excretion require water – on top of what our strength training demands. Clearly, adequate water intake should be on our minds each and every day.

Water (or as it is known chemically, H_2O), is considered the most indispensable nutrient, in that we will perish without it more quickly than if we are deprived of food. It is used by the body in many ways, and is present in all body cells – the brain, heart, skin, stomach, biceps, and even hair. A key wake-up call in this handbook has been the loss of body muscle (sarcopenia) as we age. Well, we also must realize that even more critical is the fact of daily water depletion to the body (**dehydration**). The effects of dehydration due to zero water intake, if not addressed, result in death within days to a couple of weeks. Before death occurs, the person may suffer from kidney and other organ failure, swelling of the brain, seizures, fatigue, and go into shock.

▲ **Figure 5.9** In addition to protein intake, water intake is a big part of strength training and muscle building. Muscles need to become hydrated or they will shrivel and not look full and strong. Nor will they be able to perform at peak output. Without enough water, we will not make gains in our training in spite of the fact that we may be working out consistently!

Yes, we lose water – big time – on a daily basis. In fact, we are partially dehydrated every morning when we arise, because we have just spent somewhere around seven-plus hours sleeping. This is why drinking a lot of water each day is essential – both to replace losses, and for specific benefits. Water losses occur through perspiration, respiration, urination, and defecation. To counteract just these, we should drink not only when thirsty, but push in a extra water each day. At minimum, it makes sense to drink water immediately upon arising, and at every meal (example: three times) during the day, as well as before bedtime. However, the fact that we are strength training, research has suggested that we aim for a substantial intake.

While advice for older men ranges far and wide, one is to aim for a minimum of one or if possible, two gallons per day (depending on your body weight and amount of strenuous exertion). In addition, some readers may have heard of the 8x8 rule, which means eight 8-ounce glasses of water each per day. Is this enough? Since 8 fluid ounces is equal to one cup, and there are 16 cups in a gallon, 8 glasses of water is only a half a gallon – so, it might be better to apply an 8x16 rule, which provides a full gallon – or double that one to an 8x32, to realize two gallons.

The manner in which we intake water is also important. Rather than gulping down, say, a half-gallon all at once, it is better to drink at a controlled pace over time. Also, we definitely need water before, during, and after each workout. So: water upon rising, water at meals, water throughout the day, water before, during and after workouts, and water upon retiring. This is a healthy lifestyle routine to get in the habit of doing – in spite of the minor inconvenience of more bathroom trips. Yet so many people neglect water intake in favor of other beverages, which although they contain water, as do many foods, these do not rehydrate body cells as effectively as water does. One exception: chocolate milk!

PROTEIN INTAKE

Readers of this handbook are already aware of the many benefits to consuming adequate protein, beyond building muscle. Research shows that older adults who consume adequate protein will not lose functioning the way those who do not will. But we must never become so caught up in a kind of "protein hysteria" that we neglect the rest of the nutrients the body needs. Protein alone is insufficient to sustain us, important though that it is.

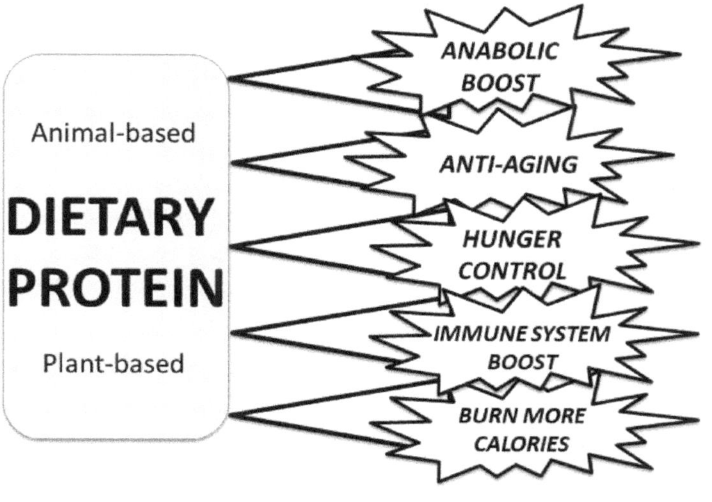

▲ **Figure 5.10** As a reminder, we are fighting against multiple challenges as we age. Research is examining the positive role that extra dietary protein for seniors can have in this regard. A high protein diet is an anabolic stimulus, essential to building muscle when combined with our workouts. In turn, having more lean body mass will burn more calories throughout the day and overnight, which promotes weight loss. Protein has a higher satiety value than fat or carbohydrate, which decreases the hunger hormone ghrelin and our desire to eat. Protein also activates immunological mechanisms.

How much protein?

Currently, 0.8 grams of protein per kilogram of body weight (0.36 g/lb) is considered the minimum amount for a healthy adult. However, there is a growing body of research evidence suggesting that this amount is insufficient for seniors, considering anabolic resistance, sarcopenia, and if the individual is strength training. The minimum requirement may be 1.0–1.2 g/kg, and higher amounts up to 2 g/kg or even more may be optimal, according to some researchers. Note - if you are obese (have a BMI greater than 30), you should calculate your protein requirements based on your goal or ideal non-obese body weight rather than your current obese body weight, so as to avoid overeating.

VARIOUS SUGGESTED DAILY PROTEIN INTAKES				
BODYWEIGHT	**0.8 g/kg**	**1.2 g/kg**	**2.0 g/kg**	**1.0 g/lb**
100 lbs.	36 g	54 g	91 g	100 g
125 lbs.	46 g	68 g	113 g	125 g
150 lbs.	54 g	82 g	136 g	150 g
175 lbs.	64 g	95 g	159 g	175 g
200 lbs.	72 g	108 g	182 g	200 g
225 lbs.	82 g	122g	204 g	225 g
250 lbs.	91 g	136 g	227 g	250 g

▲ **Table 5.2** The standard RDA suggested protein intake of 0.8 g/kg of body weight compared to more recent recommendations based upon research, for older men weighing between 100 and 250 lbs. The column to the far right, which suggests one gram of protein per pound of body weight, has been the traditional practice of bodybuilders for many years. Such an amount can be challenging to achieve through food intake alone, and may require a supplemental beverage such as a whey protein beverage. (Adapted from Schoenfeld and Aragon, 2018).

There are a couple of other things to keep in mind regarding protein intake. One is, how much is optimal at each meal, and the other is regarding the timing of protein consumption on workout days. The latter issue means to understand any benefits to having a pre-workout protein meal versus one that happens after the workout. In the past it was suggested that 20-30 grams of protein be consumed at each meal, with approximately 3-5 meals spread throughout the day. More recent research shows that the optimum amount may be 50 grams of protein at each meal.

Regarding the timing of protein meal consumption, it has been suggested that an "anabolic window" exists, and that protein should be consumed soon (example: within 2 hours) after the exercise session finishes in order to best stimulate MPS. You may recall that we covered nutrient timing back in Chapter 3. However, recent studies throw into question the need for a specific timing linked to such a limited opportunity window. The total amount of protein consumed is possibly a more critical factor than the timing. As long as the protein is ingested within a broader range of time (several hours) following exercise, the desired muscular adaptations (strength and size increase) should occur.

More studies in older men are warranted, however, in order to confirm such suggestions for our age group. The best strategy for us at present is to try for up to 50 grams of protein at each meal, and to be sure to consume one such meal within several hours of our workouts.

We should also realize that consuming increased amounts of protein is important not only on training days, but every day. Remember – muscles do not grow in response to the dual anabolic stimuli of protein intake and strength training immediately after the workout. The process occurs during the time (approximately 12-24 hours) between workouts.

Another point is regarding any possible dangers: is all this dietary protein is bad thing for the kidneys? The answer from the science at present, based upon much evidence, is that unless the person has a pre-existing kidney condition, diets high in protein do not negatively pose any problems to the kidneys or elsewhere in healthy individuals.

Which protein sources should we be consuming?

Dietary protein can come in one of two forms, complete protein foods (providing all 9 essential amino acids) or incomplete protein foods (deficient in one or more). Complete proteins contain all essential amino acids (EAAs) needed for MPS, which means they are fine to take by themselves. Complete proteins are typically derived from animal and dairy sources. Lean red meats, poultry such as chicken and turkey, fish such as salmon and tuna, eggs, and dairy products such as cottage cheese, milk, and yogurt.

There are a couple of good plant sources of complete protein, soy (think tofu and soy protein powder), and the grain called quinoa. Other vegetable-based proteins are often classified as incomplete proteins, meaning the protein is missing one or more essential amino acids and therefore would need to be combined with a second protein source missing a different amino acid. This would then provide all essential amino acids needed for optimal MPS, and is a practice called complementation or mutual supplementation. Beans (deficient in methionine but containing the other eight EAAs) combined with rice (deficient in lysine but containing the other eight EAAs) is the classic example, as is a peanut butter sandwich on whole grain bread.

After soy and quinoa, some of the more popular vegetarian or vegan options include: chia seed, beans, peas, lentils, rice, amaranth and hemp protein. The "new kid on the block", although it has been around for thousands of years, is **moringa** (Moringa oleifera). This is a tree native to India, whose leaves and pods are often eaten ground up similar to a powdered flour or meal. Having both antioxidant and anti-inflammatory properties, it has also been reported that moringa may contain anti-nutrients that reduce the absorption of dietary minerals and protein. Therefore, it is probably best to consume moringa at a time of day separate from your main protein meals.

Comparing costs of protein foods

Since there are many protein foods potentially available to the consumer, and since they vary widely in cost, it would be useful to have information to help make informed purchasing decisions. Therefore, I have examined several protein foods, calculating their cost per gram of protein.

So let's consider the list of protein foods that I have put together. It is meant to be information that any reader can consider, and put to good use right away. I've broken things down to show how much protein you are getting per serving, and that will translate to the comparative cost for an identical amount of protein in each item. For example, which food provides more protein per serving: Greek yogurt, deli ham, or a glass of milk? And beyond that - which food is most cost effective in terms of cost per gram of protein?

Based upon calculations (shown in part in the table), the ranking from most/high to least/ low in terms of protein content per g of serving is:

Greek yogurt (17 g) > 2% Fairlife milk (13g) > deli ham (12 g).

However, in terms of cost per gram of protein, the situation is:

deli ham (8 cents) > Greek yogurt (4 cents) > 2% Fairlife milk

Summarizing for these three foods:

- Each is a good source of high quality (animal-based) protein, providing between 12-17 grams per serving.
- Of the three foods analyzed, 2% Fairlife milk represented the best value in terms of providing the lowest cost per gram of protein in the product.
- Each of these foods could fit into a daily meal plan as a protein food. Milk might go well with cereal at breakfast; the ham could be part of a sandwich at lunch; and the yogurt could serve as a dinner dessert.
- A person consuming one serving of each of these three foods per day would consume (17+12+13) = 42 grams of protein. Two servings of each would provide double (84 g) the protein.
- A person would need to consume other foods of course, including plant-based foods as part of a balanced diet in addition to these in order to receive all the calories they needed plus the other nutrients lacking in these foods.

FOOD ITEM	COST	SERVING SIZE	PROTEIN (g) per serving	COST per g protein
GREEK YOGURT (32 oz., 907 g)	\$3.47	¾ cup (181 g)	17	4 cents
LOWFAT DELI HAM (9 OZ, 255 g)	\$4.28	2 oz. (57 g)	12	8 cents
FAIRLIFE 2% MILK (52 oz., 1538 g)	\$3.18	1 cup	13	3 ½ cents

▲ **Table 5.3a** Table of cost comparisons between three protein foods. For the Greek yogurt, there are 17 grams of protein in one serving (181 grams) of product, and the label indicates 5 servings total. Calculation shows that 69 cents gets you 17 grams of protein in this product, which comes out to (69/17) = approximate cost of \$0.04/gram of protein. Similarly, for the low-fat deli ham, 95 cents gets you 12 grams of protein, which comes out to (95/12) = about \$0.08/gram of protein. And for the Fairlife 2% milk, 45 cents gets you 13 grams of protein, which comes out to (45/13) = about \$0.035/gram of protein.

Some other protein foods that I examined are as follows: eggs, cottage cheese, packaged tuna fish, frozen meals ("Power Bowl", "Protein Bowl"), baked beans, and breakfast cereal (Special K Protein). The following list shows the costs per gram of protein in the dozen foods analyzed:

FOOD ITEM	COST	SERVING SIZE	PROTEIN (g) per serving	COST per g protein
KELLOGG'S SPECIAL K PROTEIN CEREAL (19 oz., 539 g)	$3.88	1 ½ cup (59 g)	15	3 cents
HEALTHY CHOICE POWER BOWL (9.75 oz. , 276 g)	$3.48	9.75 oz. (276 g)	26	13 cents
COTTAGE CHEESE Low-fat (24 oz., 680 g)	$1.64	½ cup (110 g)	13	2 cents

▲ **Table 5.3b** Table of cost comparisons between three protein foods. Notably low in cost and high in protein are cottage cheese and Special K protein breakfast cereal. However the power bowl packs quite a protein punch, and combines a variety of food groups into a healthy processed food meal that is low in calories, fat, and sodium but high in protein, fiber and which offers the convenience factor too (microwave ready in three and a half minutes).

FOOD ITEM	COST	SERVING SIZE	PROTEIN (g) per serving	COST per g protein
LARGE EGGS dozen (27 oz., 765 g)	$1.20	1 egg (57 g)	6	1.5 cents
TUNA INFUSIONS (2.8 oz. I 80 g)	$1.92	1 cup (80 g)	22	13 cents
BUSH'S BAKED BEANS (Original) (28 oz., 794 g)	$1.97	28 oz. (794 g)	7	4 cents

▲ **Table 5.3c** Table of cost comparisons between three protein foods. Eggs represent a truly economical way to buy high quality protein. Although the least expensive protein source, as you can see, one egg is a less concentrated source of protein compared to the tuna.

FOOD ITEM	COST	SERVING SIZE	PROTEIN (g) per serving	COST per g protein
GREEN GIANT PROTEIN BOWL (10 oz., 283 g)	$3.47	10 oz. (283 g)	13	27 cents
FLAT OUT MULTI-GRAIN WRAP (11.2 oz. , 318 g)	$2.98	2 oz. (57 g)	7	7 cents
EQUATE MILK PROTEIN SHAKE (11 fl. oz., 312 g)	$1.25	11 fl. oz. (312 g)	30	4 cents

▲ **Table 5.3d** Table of cost comparisons between three protein foods. For the Greek yogurt, there are 17 grams of protein in one serving (181 grams) of product, and the label indicates 5 servings total. Calculation shows that 69 cents gets you 17 grams of protein in this product, which comes out to (69/17) = approximate cost of $0.04/gram of protein. Similarly, for the low-fat deli ham, 95 cents gets you 12 grams of protein, which comes out to (95/12) = about $0.08/gram of protein. And for the Fairlife 2% milk, 45 cents gets you 13 grams of protein, which comes out to (45/13) = about $0.035/gram of protein.

Some other protein foods that I examined are as follows: eggs, cottage cheese, packaged tuna fish, frozen meals ("Power Bowl", "Protein Bowl"), baked beans, and breakfast cereal (Special K Protein). The following list shows the costs per gram of protein in the dozen foods analyzed:

COST PER GRAM OF PROTEIN OF VARIOUS FOODS

1½¢	2¢	3¢	4¢	7¢	8¢	9¢	13¢		27¢
Large Eggs	Cottage Cheese		Greek Yogurt		Low-fat Deli Ham		Healthy Choice Power Bowl		Green Giant Protein Bowl
		Fairlike Milk		Flat out Multigrain Wrap					
		Kellogg's Special K					Tuna Infusions		
			Bush's Baked Beans						
			Equate Milk Protein Shake						
LEAST ←									→ MOST

▲ **Figure 5.11** Summary of cost from least to most, left to right, showing cost per gram of protein in twelve protein food examples. As shown, eggs are the most cost effective way to consume protein of all these protein food examples. Since it is important to provide variety in the diet, any of the other foods in addition to eggs, can contribute protein. And while the emphasis has been on protein in this section of the chapter, we need not forget that each of these foods will contribute their own unique array of other nutrients as well.

Which Supplements Should I Consider?

During my training as a nutritionist, it was generally felt that the best way for the average adult to obtain all the roughly 50 nutrients that the body requires was from whole foods rather than supplements. This sentiment held true during my tenure with the USDA nutrition assistance programs as well (years 2001-2004). However, based upon more recent research and the fact that we are dealing with a more specific segment of the population having unique needs – senior men – I will suggest several supplements that appear to have at least some research evidence to warrant their use for specific purposes. Even so, I wish to emphasize at the outset that it is generally better to consume the plant foods from which substances that become supplement industry stars originate, rather than the supplements themselves. The supplement industry is a huge multi-billion dollar industry, with profit as a primary motive.

SUPPLEMENTS DISCLAIMER: The supplements discussed below are included because of historic use in connection with potential benefits, and most have at minimum been assessed via either a model system, in vitro, or in vivo research study protocol. Supplement research looking at improvements in health and functioning fundamentally seeks to gain an understanding of cause and effect as evidenced by statistics testing, although depending upon the study design and other factors, this may not always be possible. In such instances, associations (weak, moderate or strong) rather than cause and effect may be concluded. Studies may vary from in vitro cell studies seeking to understand potential mechanisms of action, to in vivo animal and human studies. Specific outcomes in cell and animal studies do not necessarily translate into guaranteed results in humans. Also in the human realm, are epidemiological and anecdotal studies. The former looks at trends in relevant populations from which general conclusions may be drawn, while the latter, although interesting to review, do not constitute a valid scientific argument.

In human clinical trials (the most desirable of which is the randomized, double-blind crossover placebo controlled study), sometimes hard data that substantiate cause and effect is lacking, and we are left with only potential rather than proven associations between supplement and outcome. There is also what is called "the placebo effect" which has been likened to "mind over matter". This basically means that a sugar pill or other inactive substance, when administered to compare with the test supplement, results in a measured benefit because the person taking the placebo "expects" or "believes" there will be a favorable response to it. It is thought that this effect is not limited to the placebo but can also carry over to the test substance (the supplement) itself.

Of all the supplements listed, among the most heavily researched are creatine, whey protein, Vitamin D, and fish oil. Due to this fact and the documented results that have been published, these might be considered as core supplements for the older man, compared to the others, although even with these there can be disagreement in the scientific literature. Realize too, that the health status of the person plays a key role. For example, Vitamin D supplementation has been shown to generate benefits primarily when the individual has a deficiency of this vitamin. This rule may apply to other supplements discussed in this handbook as well. Also, it bears mentioning that there are individual differences in results to supplement use – meaning that your friend may experience benefits with "supplement X", and yet you do not.

Many supplements still require further research before they can receive a solid recommendation. Sufficient data is lacking in order to draw definitive conclusions. Also, many plant bioactives lack long-term safety evaluation. Regarding supplements I have listed as being "anabolic", it is important to realize that any muscle growth effects would be minimal when compared to that achieved via anabolic steroids, or to the practice of consistent resistance training combined with extra dietary protein. Thus, supplements are supplemental in the truest sense of the word.

So we must approach the following discussion of supplements with eyes wide open – willing to acknowledge a possible lack of effectiveness as well as potential risks and side effects, however minimal they may be. We ought to also realize that supplements should be viewed as secondary lifestyle strategies when compared to the primary strategies of regular exercise, sufficient protein and water intake, proper calorie intake to achieve and maintain a healthy weight, proper distribution of **macronutrients** with an emphasis on plant-based foods and healthy animal based foods, and adequate rest and sleep.

SUPPLEMENT	ANTIOXIDANT	ANTI-INFLAMMATORY
Ashwagandha	✓	
Astaxanthin	✓	✓
Curcumin	✓	✓
Fiber		✓
Fish Oil		✓
Green Tea		✓
Minerals	✓	✓
NSAIDs		✓
Quercetin	✓	✓
Vitamin D		✓

▲ **Table 5.4** There are numerous supplements available to the consumer, so many in fact that it is overwhelming. Not all have proven benefits, and not many are necessary for older men to consider. However, the ones listed in this table which are specifically categorized as antioxidants or anti-inflammatory agents are discussed in this section on supplements, which I recommend that you read over carefully. For our purposes, it is important to remember that chronic inflammation not only contributes to aging and age-related disease onset, but also interferes with muscle and strength development.

Most of the substances that are suggested as supplements are plant-derived. If a news report states that substance x has been associated with a positive health outcome in a new study, this often leads to the two h's: "hope" followed by media "hype". What really is needed, and often is lacking, is what are called randomized double-blind placebo-controlled research trials using people. These represent the gold standard for determining whether a supplement has a therapeutic (disease prevention or health enhancement) potential. Such trials need to be repeated in diverse groups, and followed long-term. Any drug interactions and adverse effects need to be fully understood. This makes such trials very expensive and often impractical. In many cases, the only data available for a particular substance is from cell studies (called in vitro research). Lacking human clinical trials, animal model studies (rodents, dogs, etc.) are often conducted as part of in vivo research. At times, there are only a handful of studies from which to draw any (very tentative) conclusions.

The main classes of supplements for us will be (1) those that assist with building muscle – hence, are anabolic or anti-catabolic; (2) those that help restore more youthful hormone levels; (3) those that are antioxidants and therefore fight oxidative stress and aging; and (4) those that are anti-inflammatory and therefore fight chronic inflammation and inflammaging.

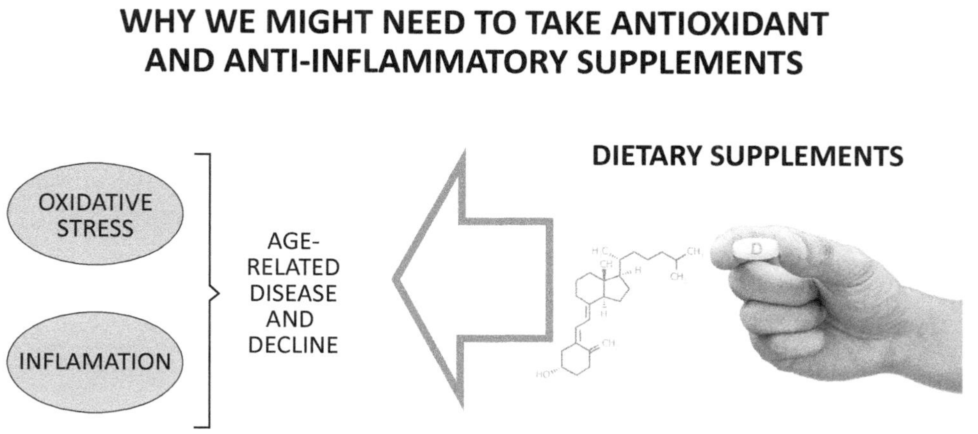

▲ **Figure 5.12** Recall that oxidative stress at the cellular level is one of the suggested causes of aging. Also recall that chronic inflammation (LGI, low grade inflammation, or inflammaging) is thought to be a main cause of many diseases that befall us as we age (cancer, diabetes, obesity). In addition, inflammation can inhibit MPS (muscle protein synthesis), which would then negate our strength training and anabolic nutrition lifestyle. This suggests a role for supplements that are antioxidant, anti-inflammatory, and anabolic.

As has been said previously, ingesting protein and performing resistance exercise training are the two primary anabolic activities that we can undertake to develop (rather than lose) muscle and strength. However, I will now provide a discussion of supplements that older men can consider taking on a daily basis (not just on training days) in order to address aging, sarcopenia, anabolic resistance, and MPS. Generally speaking, we will focus on supplements that enhance anabolism and are anti-aging to some degree. But it should also be obvious that if there is a way to slow down protein and muscle catabolism, this would also prove beneficial. This is because we would lose less muscle mass in a given period of time, which means that by difference, we would gain more muscle "net" through anabolic practices. What I am trying to say is, we should consider adding to our daily regimen any direct anabolic supplements, plus any anti-catabolic substances (these are indirectly anabolic). Although these are the first potential substances to be discussed, let me make it clear at the start that left out of the equation will be anabolic steroids.

Anabolic steroids

Anabolic steroids are man-made analogs of the male sex hormone testosterone, taken in order to promote muscle mass. These should be only available and used legally through a doctor's prescription, and based upon a proven medical need as diagnosed by your physician. They are to be used under the doctor's supervision due to potential side effects. The doctor will order routine blood analyses to be sure that adverse effects are not developing, and if they are, will reduce the dose or stop use, and consider alternative therapies. It is interesting that strength training protocols when using anabolic steroids are different from what the rest of us (who train "naturally", meaning without steroids) require. The anabolic response is much greater, with significantly less muscle overload required, in a steroid user. Stated in humorously exaggerated terms, "a person using steroids merely has to think about lifting weights in order to see massive muscle growth."

Ashwagandha (Withania somnifera)

The medicinal herb **ashwagandha** has been used for centuries for a variety of purposes. These include increasing energy levels and reducing stress. As a supplement for older men, it is now available in a potency of up to 1300 mg capsules. This substance is classified as an **adaptogen**, which means that when taken as recommended, it can alleviate the negative effects of exercise and stress, including fatigue, depression, and anxiety. According to some studies, ashwagandha may also have anabolic properties, and may aid in MPS.

SUPPLEMENT	ANABOLIC	ANTI-CATABOLIC	OTHER
Ashwagandha	✓		✓
Astaxanthin	✓		✓
Beta alanine			✓
Creatine	✓		
Curcumin	✓	✓	
DHEA	✓		✓
Epicatechin		✓	
Fish Oil	✓		✓
HMB	✓	✓	
Leucine	✓		✓
Quercetin	✓		✓
Vitamin D	✓		✓
Whey protein	✓		

▲ **Table 5.5** The supplements listed in this table have been reported to elicit either anabolic, anti-catabolic, and in certain cases, both effects. Any of these might be worthy of consideration in conjunction with the Old Man Muscle workout, a sound diet, plus adequate rest and sleep.

Astaxanthin

Astaxanthin is a pigment molecule in the carotenoid family (think of beta carotene). It occurs in the animal kingdom (salmon) as well as plant kingdom (marine algae). As a supplement, it is beginning to be more widely studied, and is available as 10 mg capsules. One of the interesting things about this substance is that it displays both antioxidant and anti-inflammatory properties. It has also been formulated into an anti-sarcopenia drug called AstaMed MYO™, which also contains Vitamin D3. The manufacturer states "nutritional supplementation with the astaxanthin formula has been shown in clinical studies to maintain and support optimal muscle size and function. AstaMed MYO™ provides a clinical solution. In short, diet, exercise plus AstaMed MYO™ is a new protocol for sarcopenia."

Beta alanine

Beta alanine is a non-essential amino acid that has been shown to enhance exercise performance, specifically strength gains. Reportedly, this supplement increases the production of carnosine synthesis in muscle. This is important because the dipeptide carnosine has definite proven effects on exercise performance through its buffering (maintaining a stable pH) ability. This means it can lessen the painful and fatiguing impact of lactic acid build-up in muscle, which is generated during intense exercise. Studies used doses of 6.4 grams per day, in eight divided doses of 800 mg each. Be advised, it is primarily effective only when you exercise at high intensity.

Creatine

Creatine (marketed as creatine monohydrate powder) is a relatively inexpensive yet bioactive substance that has been exhaustively researched for safety and effectiveness. Creatine is stored in skeletal muscle, and is called upon to generate ATP. Studies have also recently linked it to muscle hypertrophy, and possibly reducing muscle breakdown (anti-catabolic effect). While a small percentage of users will be "non-responders" (will not see results), most older men may benefit from this supplement. I suggest taking 2 grams once or at most twice a day, each time with a meal high in both protein and carbohydrate. This can be at a mealtime of your choosing (breakfast, lunch or dinner) and also just before or after your workout. There is no need to practice creatine "loading" because taking this supplement daily for a few weeks will increase your muscle creatine levels to the same extent as if you loaded initially with 20 grams.

Curcumin (turmeric)

Curcumin is a phenol compound which is a component of the spice called turmeric. Phenol compounds are recognized due to their molecular six-membered ring structures having hydroxyl (-OH) groups attached to the ring(s). Research has indicated that, similar to astaxanthin, curcumin not only acts in an anti-inflammation manner but also as an antioxidant. In protein wasting conditions associated with cancer (cachexia), it appears that curcumin may also be slightly anti-catabolic, meaning it may help the body hold onto muscle. It may also boost the anabolic effect of insulin by increasing insulin sensitivity. You can find it in 1000- 2000 mg capsules.

DHEA

This substance, **dehydroepiandrosterone (DHEA)**, is a prohormone secreted by the adrenal glands that is enzymatically converted into the primary **hormones** estrogen and testosterone in the body. It has been touted as a "fountain of youth" supplement by promoters, due to its alleged ability to restore hormones to youthful levels. However, the research is lacking to back up such a sensational claim. It is known that DHEA levels do steadily decline as we age, and when supplemented in the **diet** (for men, 5 to 50 mg per day is suggested), effects may include a slight boost in testosterone. There has been a suggestion of an anabolic role for DHEA as well as being able to improve mental sharpness, but studies are needed to substantiate such claims. Some studies using higher does (100 – 200 mg per day or more) have reported acne in users, hair loss, and the possibility of an irregular heartbeat. Due to DHEA's interaction with hormones, it may be unwise for anyone with prostate cancer to use this supplement, as it may stimulate tumor growth.

Epicatechin

Research clearly indicates the negative effects that aging has on body regulators of muscle growth and differentiation. However, consumption of **epicatechin**, which is a green tea flavanol, seems able to somewhat reverse these changes. This points to a role for it in the treatment of sarcopenia, and possibly in muscle building. We learned a bit about satellite cells and myogenesis in the exercise science chapter. We appreciate too that the mTOR signaling pathway has a positive effect on muscle growth, therefore is anabolic. On the other side of the equation is **myostatin**, which has the opposite (catabolic) effect: muscle growth is inhibited due to its action. The interesting thing is, epicatechin has been shown to inhibit myostatin, which means that epicatechin is an anti-catabolic substance. This supplement is available in 300 mg capsules.

Fish oil

Fish oil, which contains omega-3 fatty acids (DHA and EPA) that act in an anti-inflammatory manner, has been shown to promote muscle protein anabolism. Regarding how fish oil works, it is believed that omega-3 supplementation can increase muscle protein synthesis by regulating the mTOR pathway, which is a very important regulatory pathway for cell production and growth. Omega-3 fatty acids in fish oil act in an anti-inflammatory manner by inhibiting the pro-inflammatory molecule called NF-*k*B. This reduces muscle protein catabolism. In addition, it is believed that fish oil (specifically, the EPA fatty acid) improves anabolism by enhancing sensitivity to anabolic stimuli including exercise, protein intake and increases in insulin levels (the latter two being nutritional signals).

HMB Beta-hydroxy-beta-methylbutyrate

Beta-hydroxy beta-methylbutyrate (HMB) is a derivative of the amino acid leucine that is thought to be anti-catabolic (prevents muscle breakdown). HMB is believed to assist with muscle hypertrophy and strength development as well. It is also believed to decrease muscle recovery time following a workout. HMB supplements are available as 1500 mg capsules.

Leucine

Leucine is one of the BCAAs (branched chain amino acids) and is the primary amino acid to trigger MPS. In studies assessing leucine's effect on muscle protein synthesis, it appears that leucine upregulates MPS. However, leucine may be be more effective at promoting gains in muscle in those with lower dietary protein intake and also in those having anabolic resistance to protein (i.e. the elderly). It is available in 1500 mg capsules, and is also a major component of whey protein powders.

NSAIDs

Most readers are familiar with **NSAIDs**, or non-steroidal anti-inflammatory drugs like acetaminophen, ibuprofen and aspirin. Over-the-counter NSAIDs remain a popular treatment strategy to address exercise-induced inflammation and pain. They work by inhibiting what are called the cyclooxygenase (COX-1 and 2) enzymes, preventing the formation of prostanoids (prostaglandins, prostacyclins, and thromboxanes), which are compounds produced in the body that cause acute inflammation. Inflammation is associated with blunted anabolic response and enhanced catabolism in the elderly. A few studies have investigated the use of NSAIDs to reverse these effects. There has been a report that NSAIDs (4 g per day acetaminophen, or 1.2 g per day ibuprofen) promote both muscle hypertrophy and strength gains in the elderly who resistance train. As a result it is tempting to suggest the proactive daily consumption of NSAIDs to overcome age-related muscle anabolic resistance. However the associated risks, such as cardiovascular, kidney, liver, and stomach adverse effects, make that a poor course of action. Additionally, there is a suspicion that the COX-2 enzyme actually plays a positive role in muscle hypertrophy, possibly in relation to satellite cell activity. Having said all of this, realize that NSAIDs do remain recommendable on occasion to alleviate DOMs and the pain of minor activity-associated injuries. It is chronic, long-term NSAID use, especially at sustained higher doses, that is to be discouraged.

Quercetin

Quercetin is a bioflavonoid molecule found in a great variety of foods, including apples, berries, broccoli, and tea. It has been reported that this substance has anti-inflammatory, antioxidant, anticancer, antihypertensive, and blood glucose-lowering properties. Quercetin works by scavenging reactive oxygen species (ROS) and by targeting prominent pro-inflammatory signaling pathways (STAT1, NF*k*B, and MAPK). As with many supplements however, there is a discrepancy between results obtain in vitro versus those obtained in vivo, which may be explained in part due to low bioavailability when the supplements are ingested by mouth. A modified isoquercetin supplement (EMIQ) that has improved bioavailability has been shown in a 2019 Japanese study to increase muscle protein synthesis. This means that the enhanced quercetin form displays antioxidant, anti-inflammatory, and anabolic potential. The study involved teenage athletes; no research has yet been performed using isoquercetin in older men. Standard (not isoquercetin) capsules are available in the 500-1,000 mg range.

Sinensetin

I must say, I found it rather humorous that there would be a substance, the citrus plant flavonoid **sinensetin**, that is abbreviated as SIN. Which begs the question, is SIN good for us? Okay, I've had my fun. This compound actually came to my attention only hours before submitting the manuscript of this book to the publisher – so I have rushed to include it. And I must apologize – it is not actually available as a supplement. At least not yet. It is in the family of plant PMFs (polymethoxyflavones), and has exhibited anti-inflammatory, antioxidant, and anti-cancer properties. In a recent (2019) study using rats, investigators hypothesized that SIN might be able to prevent sarcopenia and improve the function of muscle satellite cells – for the time being, only in rats. My goodness... So, for anyone interested, I discovered that orange juice contains 0.1-0.2 mg SIN/100 ml of juice – not really much, unfortunately.

Vitamin D

You might recall learning about the "sunshine vitamin" years ago in school. What? No, it is not Vitamin C - I think you must be associating sunshine with Florida oranges! **Vitamin D** is called the sunshine vitamin because when the skin is exposed to sunlight, the body synthesizes this vitamin. However, many older persons for a variety of reasons (including a reduced ability to synthesize it) are deficient in this vitamin, the active form of which is Vitamin D3 (cholecalciferol). We can turn to foods as a source, the best of which would be fatty fish, fortified breakfast cereals, milk, and milk products like yogurt. These dairy products are probably the best choices because they also contain calcium and are easy to consume every day. One of the main functions of Vitamin D is to enhance calcium absorption in bone to prevent osteoporosis. Recently, this vitamin has also been shown to play a regulatory role in metabolic pathways implicated in muscle wasting, and in the regulation of the immune system. Vitamin D supplementation in the elderly has been shown to enhance muscle strength but not muscle mass. The way it works is under investigation and appears to be related to the anabolic signaling of insulin as well as that of leucine on MPS. This vitamin has also been studied with respect to boosting testosterone, and seems to have a positive effect in older men who have testosterone deficiency. As a daily supplement, 2,000–3,000 IU (international units) of Vitamin D3 may be taken with a meal.

Whey protein

Whey protein has gained fame as the food protein with the highest biological value, even exceeding that of egg proteins. Whey is one of the two major proteins in milk, in addition to casein. Whey protein provides significant amounts of the essential and branched chain amino acids to promote MPS. Whey protein digests more quickly than casein and has a higher concentration of leucine, the most potent amino acid to stimulate MPS to help overcome anabolic resistance. Whey is considered a fast-acting protein, while casein is slow-acting (to stimulate MPS). This is why whey is a top choice right after a workout, while casein is good to consume overnight during recovery. The slow release of amino acids provided by casein taken at bedtime (a glass of milk) is a great anti-catabolic strategy. Whey protein can be purchased as a whey protein concentrate protein powder, but is also sold pre-blended into protein beverages for convenience. A whey beverage should be viewed as a supplement, and not serve as the primary means for consuming dietary protein throughout the day – that task should be addressed through the consumption of protein-rich foods. However, due to the fact that we are needing to increase our protein intake as older men, a whey shake here and there, providing 30 grams of protein, in combination with protein foods, will help us to more easily reach our daily protein goals.

What about supplements to boost testosterone?

First things first: if you suspect low testosterone (symptoms include being generally tired, lacking mental sharpness, experiencing decreased sexual drive, having more having body fatness than body muscle), it makes sense to visit your doctor to get a testosterone check. This involves a simple blood analysis for total and free testosterone. The doctor has the ability to prescribe testosterone hormone therapy if warranted.

Now - one of the aging challenges mentioned earlier in this handbook was that of declining testosterone. With testosterone decline comes decreases in energy and mental sharpness, reduced ability to build muscle, depressed mood, and decreased libido, to name just a few of this key hormone's functionalities. So naturally, older men are rightfully interested in knowing if there are any ways to stimulate its production. I have mentioned that habitual strength training that employs multi-joint, multi-muscle group involvement such as squats, deadlifts, and thrusters (but let's not omit flipping heavy tires or pushing heavy weight sleds!) have been shown to ramp up T levels. Also, it may be about increasing training volume as well – that which influences muscle hypertrophy. But avoid endurance training – this is catabolic, and reduces our T levels. Leisure activities like walking and bike riding are certainly fine, but regular high intensity training on a bike or marathon running will work against T levels. Medium to high intensity weight training is different, and is anabolic and pro-testosterone.

A big issue with testosterone decline is the presence of belly fat. I say this because there appears to be an association between the amount of belly fat a man has and the reduction in T levels. This is due to the conversion of testosterone, which we know is the primary male sex hormone, into the primary female sex hormone, estrogen, which is stimulated by the presence of belly fay. Yes – you read that right. So it becomes important to lose that belly if you want higher T levels. By the way, a man with robust T levels will burn fat more efficiently than one who has low T, which means it will be easier to maintain a lean and muscular physique with higher T levels.

As far as foods go, it has been proclaimed on the internet that beet root, celery, and white button mushrooms will help, although I could not find any well-designed studies in the scientific literature to substantiate these claims. Adding to this is the multi-million dollar **testosterone booster** supplement industry, which unfortunately also lacks definitive and properly designed and executed human clinical trials as evidence for efficacy. Typical supplements contain herbs such as maca root, fenugreek, longjack (Eurycoma longifolia), Tribulus terrestris, or a "proprietary blend". While these may work to boost libido, which can be desirable, this is not the same thing as boosting testosterone itself.

The word "boron" may sound like a cross between "boring" and "moron", but this is an exciting little substance. There may be several important metabolic roles for the mineral boron, which has been reported to increase the conversion of total testosterone in the body to the active free testosterone form. It also may help Vitamin D with boosting T levels, improve brain function, and block chronic inflammation - all of which is good news for older men. Apples, dried fruits like raisins, nuts, and coffee contain this trace mineral. Supplements of boron at 10 mg are also available. As with any mineral supplement, be careful not to exceed the recommended dose, as toxic effects and even death can result if too much is ingested.

While on the subject of minerals, let me mention magnesium and zinc, both of which are prevalent in foods, but also available as supplements. These will be mostly effective in persons having deficiencies of these minerals. What is the connection to testosterone? Well, testosterone can be bound to both albumin (not very tightly) or to sex hormone binding globulin (SHBG), tightly and unable to be released. Importantly, it is only the released or free testosterone that is bioactive. Magnesium supplementation can increase total testosterone. Also, by inhibiting what is called SHBG (sex hormone binding globulin that attaches to testosterone, making it functionally inactive), it can increase the percentage of active, free testosterone. Zinc appears to similarly be able to increase T levels. It is thought to be effective because it inhibits the aromatase enzyme that converts testosterone into excess estrogen. Less estrogen production means more T still available. However, increases due to Mg and Zn are considered modest at best. But, every little bit helps.

The last issue that I will mention in connection with testosterone is regarding so-called **endocrine disruptors**, and it is scary. Simply put, these are substances – such as the plastics that we come into contact with on a daily basis - that have been shown to disrupt the functioning of the human endocrine (hormonal) system. Think testosterone, think thyroid hormone, think overall health and functioning. Due to the extremely large number of plastics and other man-made chemicals in use worldwide, and the lack of precision detection methods for many, fully understanding the consequences of human exposure to such materials as endocrine disruptors is quite difficult. If you are having trouble knowing what this all means, let me put it to you directly: plastics are in use everywhere and exposure to and use of plastics has been shown to decrease testosterone.

Plastics are made from hydrocarbon starting material and contain other components that are used in their manufacture, such as phthalates, bisphenol A (BPA), dioxin, vinyl chloride, PCBs, and many more. The problem is not only that constant daily exposure to these substances decreases testosterone production in men (and boys), but also that many of these compounds function as **xenoestrogens** in the body – they mimic the female hormone estrogen and cause feminizing and sexual dysfunction effects in males. While these statements may seem ludicrous, there is ample research to back them up.

If you take a moment and conduct a mental inventory of your daily routine, you will quickly realize how frequently you come in contact with plastics. Why, there's the bottled water that you hold so dear that puts plastic in contact with the H_2O and you. Ironically, most of the supplements and whey beverages that consumers buy also are enclosed within plastic bottles and containers. Think of all those handy zip-lock storage bags being used by you and the wife to store leftover food in. Not to mention the plastic drinking cups used at junior's birthday party… and my vintage vinyl LP collection! Well, I don't eat or drink the records, so not much to worry about there. But the toothpaste tube, the toothbrush that gets put in your mouth, and plastic microwaveable containers are all culprits too. In fact, I have read warnings that the act of microwaving food in plastic is one of the most efficient and effective ways to expose the body to endocrine disruptors. At home, I now try to only store foods in glass containers, and microwave using paper plates, or glass or ceramic materials: no plastic. I'm back to using aluminum foil for storage as well. Plastics were a great breakthrough in concept – who knew that in actual practice there would be this problem? The home container and housewares industry is not likely to change practices any time soon, so it's up to us to choose the materials that we come into contact with at home and on the road as carefully as possible.

What about supplements for the brain?

You may have seen a commercial on television for a supplement made from jellyfish that is supposed to boost your brain. This supplement (Prevagen) has been exposed as total bunk in a 2019 article written by Dr. Anthony Pearson and published in JAMA. Does this mean that there are no other supplements or lifestyle changes that might be genuinely effective for memory and learning?

There are natural compounds (pigment molecules) called **anthocyanins** present especially in darker berries (blueberries and blackberries) that have a positive impact on neurological processed and may slow the age-related decline in brain and cognitive (learning and memory) functions. The anthocyanins in these berries apparently upregulate neurogenesis, neuroplasticity, and a key compound called BDNF (brain-derived neurotrophic factor). In addition to eating these fruits, supplements are available, such as blueberry extract capsules and mixed berry anthocyanin capsules. These typically provide 100 mg or more of the berry extracts. And, lest we forget, strength training has also been shown to boost brain function and BDNF levels – so combining regular anthocyanin intake with lifting is a beneficial lifestyle strategy.

We have now reached the end of our necessarily incomplete traversal of supplements. The number of supplements is truly daunting – and it is beyond the scope of this little handbook to cover them all. Apologies if I omitted one of your favorites (capsaicin, piceatannol, resveratrol, selenium, Vitamins C or E?) from the discussion. I would add that it may be prudent to take a multivitamin and multi-mineral supplement, especially if and when one's diet is poor (nutritionally inadequate). In all of this, we should realize that supplements cannot substitute for a proper diet, a regular exercise regimen, and quality rest and sleep. If our lifestyle is deficient in these primary areas, and we think that by taking supplements we will be able to "skate by", we are mistaken. Supplements by definition are just that – supplemental to what is primarily important.

SUPERFOODS

A quick word regarding so-called "superfoods" This term was coined fairly recently and refers to any food that is nutrient dense and provides health benefits. This means that just about any fruit, vegetable or grain can be called a superfood, but so also can salmon, Greek yogurt and hen's eggs. Although it is really a marketing strategy and somewhat gimmicky, the concept is not entirely without merit if it gets people to change their eating behavior more in the direction of plant foods. Remember – plants contain the greatest arsenal of antioxidant, anti-inflammation, and anti-aging compounds on the planet, and these can serve us well as older men if we add them to our diet. It would be best to make a trip to the supermarket and purchase some fruits, vegetables, and nuts. But you could also have on hand, I suppose, the powdered versions of fruits and veggies as a back-up that you can drink. Come to think of it, you could blend a tablespoon of these powders into a glass of your favorite 100% juice (e.g. apple juice, cranberry juice, orange juice, pomegranate juice, etc.), water, or unsweet tea.

▲ **Figure 5.13** Although eggs, salmon and yogurt make the superfoods list, it is a fact that superfoods by and large tend to be plant substances – such as nuts, berries, vegetables, fruits, beans, peas, sprouts, and legumes.

And are you aware of **microgreens**? These are essentially young or "baby" green plants, similar to but smaller than sprouts. They are often more nutrient dense and richer in plant bioactive molecules than their full-sized counterparts. Raising your own microgreens is fairly simple, as seed kits are available for purchase and do not take much room. The germination period is rapid, typically less than seven days. You'll be harvesting and eating more healthfully before you've had time to think about it all.

SUPERFOODS	
Acai berries	Mushrooms
Almonds	Nuts & Seeds in general
Apples	Oats
Avocados	Olive Oil
Beetroot	Pecans
Berries in general	Pomegranate
Blueberries	Salmon
Cacao (Dark Chocolate)	Spinach
Cranberries	Sprouts
Dark Leafy Greens	Sweet Potato
Eggs	Tomatoes
Garlic	Walnuts
Ginger	Yogurt
Green Tea	

▲ **Table 5.6** Table of superfoods. Though not meant to be an exhaustive list, the reader can get a good idea of the diverse variety of foods that will fill him up (in addition to the protein foods and water he will consume), leaving little or no room for junk foods. I realize that for some, this is a radical shift in one's eating habits. So, if the thought of a baked sweet potato is challenging, sprinkle some brown sugar and add a pat of butter. If dark leafy greens like kale and spinach, with a handful of chopped white button mushrooms mixed in, is repulsive, go ahead and add a helping of that Southwest salad dressing to make it all okay. If you can't stand the thought of eating a pomegranate, go ahead and start with a plate of vanilla ice cream and then mix in the pomegranate. Throw in some blueberries too, while you're at it. See? You are using a smart approach to help change your diet for the better.

Lastly, while I still advocate eating whole plant foods if possible (greens, other vegetables, fruits, nuts, seeds, legumes), the food industry has made available such new products as beet powder, cacao powder, "super greens" and "super reds", and more. These are all processed powders that are derived from the original plant material, converted into a convenient, easy-to-use form. If a person has run out of whole pomegranates, blueberries, kale, and white button mushrooms, it may be reasonable to purchase and have powdered versions of these substances on hand.

SLEEP HABITS

In Chapter 2, sleep inadequacy was mentioned as a challenge to us as we age. Both the quality and quantity of sleep is supremely important for whole body restoration to occur. Sleep has restorative effects on cognition and brain health, possibly because cerebrospinal fluid (CSF) may flush out toxic waste. This important cleaning of the brain mainly occurs during sleep. We also considered that strength training may help us get a better night's sleep by reducing stress, which also means reducing the catabolic stress hormone cortisol. One question is, "Is it better to work out in the morning, afternoon, evening – or does it really matter?"

This is actually a very important question. Many times, our externally-driven schedules dictate when we can work out and when we can sleep. Also, many of us classify ourselves as "morning people" while others consider themselves "night owls". From what I have read, here are a few suggestions:

- For many people, the demands of the day pile up and make it difficult to work out any time other than first thing in the morning. This might be 6 am for some people, or slightly earlier or later.
- People who work out early in the day tend to feel energized to face the day.
- People who work out earlier in the day tend to sleep longer at night, and have a more restorative and refreshing sleep.
- On the other hand, people who work out late (just before bed) may find it more difficult to fall asleep, and may experience a more restless sleep.
- But it is also possible that the increase in body temperature and heat generated from evening strength training workouts could relax the muscles due to a sensation similar to a warm bath effect, which coupled to fatigued muscles may contribute to a fast drop off into never-never land and a sound sleep.

These are of course generalities, and you should seek to know yourself and what works best for you. Suit your individual preference. Just realize that working out can spike the adrenaline response, and due to the exercising activities of the workout, our brain and body become energized and may take a while to wind down, relax, and allow for sleep. You can always experiment with different times of the day for workouts, and keep a "sleep log" that notes how long and the quality of your night's sleep. It is a good idea once you decide on a certain time, to stick with that time as much as possible. You are training your body clock to expect it, which may help the body adapt and make the workout part of its biological clock so that workouts become optimally efficient and effective.

If you need assistance from a natural sleep aid, you can consider using between one and three grams of a melatonin supplement just before bedtime.

STRESS

We all experience stress in our lives. Heck – just turn on the TV and watch the news for a while – you'll likely get stressed out in no time. Actually, we've mentioned the stress hormone cortisol several times in this handbook already. Cortisol levels rise in the morning and help us arise from sleep, after which they start to decline. Unless we become mentally and emotionally stressed throughout the day, that is. Higher stress levels correlate to a decrease in MPS, and this is related to decreases in anabolic hormones (growth hormone, IGF-1, and testosterone).

There are three anti-stress supplements that I will briefly mention: ashwagandha, **magnesium**, and theanine. Ashwagandha is an herb that we have discussed in this chapter that has shown some use in combatting stress in a few animal and human studies. Doses employed have varied from 125 mg to one gram or more per day. Supplementation with magnesium (typically 400 mg daily) can be a way to address stress and calm down an excited nervous system. Theanine (200 mg daily) is a naturally occurring green tea amino acid that produces effects similar to magnesium. Reductions in cortisol, the stress hormone, are typical with these anti-stress supplements.

RECAP AND WHAT'S AHEAD

We have just finished a chapter examining lifestyle behaviors that we should adopt if we are serious about thriving physically into our later years. Earlier in this handbook we first learned about sarcopenia, the debilitating age-associated progressive loss of skeletal muscle mass and strength. To add insult to injury, sarcopenia is often accompanied by an increase in body fat as a consequence of metabolic changes associated with an inactive lifestyle. What we must do is alter our lifestyle through dietary, strength training, and other applicable means to address this downward spiral. Hopefully you have come away with a new insight regarding how to best feed your body (with food as well as certain supplements) in order to build muscle, slow down aging, and reduce inflammation – which should decrease the risk of the chronic diseases common to older men – and achieve a longer, stronger life. Isn't it good to know that we can play an active role in our own health by adopting appropriate lifestyle changes? The focus was on our dietary and sleep habits, having covered physical activity (the specifics of strength training and the Old Man Muscle workout) in the previous chapter. However, there is one more piece to the lifestyle puzzle – the spiritual component – and that is the topic of the next chapter - see you there.

AHA (American Heart Association). 2015. Saturated fat. Accessed Mar. 29, 2019: heart.org/en/healthy-living/healthy-eating/eat-smart/fats/saturated-fats.

Alway SE, Pereira SL, Edens NK, Hao Y, Bennett BT. 2013. ß-Hydroxy-ß-methylbutyrate (HMB) enhances the proliferation of satellite cells in fast muscles of aged rats during recovery from disuse atrophy. Exp Gerontol. 48(9):973-984.

Amiri M, et al. 2019.Chocolate milk for recovery from exercise: a systematic review and meta-analysis of controlled clinical trials. European Journal of Clinical Nutrition. 73:835-839.

Andersen CJ. 2015. Bioactive Egg Components and Inflammation. Nutrients. 7(9):7889–7913.

Antonio J et al, 2015. A high protein diet (3.4 g/kg/d) combined with a heavy resistance training program improves body composition in healthy trained men and women – a follow up investigation. Journal of the International Society of Sports Nutrition 12:39.

Aparicio A, et al. 2018. Egg as a source of antioxidants and bioactive components against chronic processes. Nutr Hosp. 35(Spec No6):36-40.

Asai D, et al. 2000. Structural essentials of xenoestrogen dialkyl phthalates to bind to the estrogen receptors. Toxicol Lett. 118(1-2):1-8.

Berardi J. 2011. Experiments with intermittent fasting. Accessed at:http://danjohn.net/wp-content/uploads/Intermittent-Fasting_Precision-Nutrition.pdf

Bishop N, Zuniga K, Noon B. 2019. Egg Consumption and Multi-Domain Cognitive Function in a Representative Sample of Older U.S. Adults (P14-004-19). Curr Dev Nutr. Accessed at: https://www.ncbi.nlm.nih.gov/pmc/articles/PMC6573972/

Bordoni A et al. 2017. Dairy products and inflammation: A review of the clinical evidence. Crit Rev Food Sci Nutr. 57(12):2497-2525.

Bounous G, Kongshavn PA. 1982. Influence of dietary proteins on the immune system of mice. J Nutr. 112(9):1747-55.

Brouns, F. 2018. Overweight and diabetes prevention: Is a low-carbohydrate-high-fat diet recommendable? European Journal of Nutrition, 57 (4), 1301–12.

Carroll CC. 2008. Ibuprofen and acetaminophen promote muscle hypertrophy and strength gains during resistance exercise in the elderly. FASEB J. Accessed at: https://www.fasebj.org/doi/abs/10.1096/fasebj.22.1_supplement.753.31

Castro B, et al. 2013. Bisphenol A exposure during adulthood alters expression of aromatase and 5⊠-reductase isozymes in rat prostate. PLoS One 8(2):e55905.

Centers for Disease Control and Prevention. Healthy Food Environments. Accessed at: https://www.cdc.gov/obesity/strategies/healthy-food-env.html

Teixeira EM, et al. 2014. Chemical characteristics and fractionation of proteins from Moringa oleifera Lam. leaves. Food Chem.147:51-54.

Chowdhury, R., et al. 2014. Association of dietary, circulating, and supplement fatty acids with coronary risk: A systematic review and meta-analysis. Annals of Internal Medicine, 160 (6), 398–406.

Cotman CW, Engesser-Cesar C. 2002. Exercise enhances and protects brain function. Exercise and Sport Sciences Reviews 30(2):75–79.

Cruzat VF, Krause M, Newsholme P. 2014. Amino acid supplementation and impact on immune function in the context of exercise. J Int Soc Sports Nutr. 14;11(1):61.

Cruz-Jentoft AJ. 2018. Beta-Hydroxy-Beta-Methyl Butyrate (HMB): From Experimental Data to Clinical Evidence in Sarcopenia. Curr Protein Pept Sci.19(7):668-672.

Dall S, Rossmeislova L, Koppo K. 2017. The Role of Inflammation in Age-Related Sarcopenia. Front Physiol. 8:1045.

Daneshvar P, Hariri M, Ghiasvand R, et al. 2013. Effect of eight weeks of quercetin supplementation on exercise performance, muscle damage and body muscle in male badminton players. Int J Prev Med. 4(Suppl 1):S53–S57.

De Cabo R, Mattson MP. 2019. Effects of Intermittent Fasting on Health, Aging, and Disease. N Engl J Med. 381:2541-2551.

Delecroix B, Abaïdia AE, Leduc C, Dawson B, Dupont G. 2017. Curcumin and piperine supplementation and recovery following exercise induced muscle damage: a randomized controlled trial. J Sports Sci Med.16(1):147–153.

Deutz NE, Bauer JM, Barazzoni R, et al. 2014. Protein intake and exercise for optimal muscle function with aging: recommendations from the ESPEN Expert Group. Clin Nutr. 33(6):929–936.

Draganidis D, Karagounis LG, Athanailidis I, Chatzinikolaou A, Jamurtas AZ, Fatouros IG. 2016. Inflammaging and Skeletal Muscle: Can protein intake make a difference? J Nutr. 146(10):1940-1952.

Durkalec-Michalski K, Jeszka J, Podgórski T. 2017. The Effect of a 12-Week Beta-hydroxy-beta-methylbutyrate (HMB) Supplementation on Highly-Trained Combat Sports Athletes: A Randomised, Double-Blind, Placebo-Controlled Crossover Study. Nutrients 14;9(7):E753.

Etnier JL, Wideman L, et al. 2016. The effects of acute exercise on memory and brain-derived neurotrophic factor (BDNF). Journal of Sport and Exercise Psychology 38(4):331–340.

Ferreira GC, McKenna MC. 2017. L-Carnitine and Acetyl-L-carnitine roles and neuroprotection in developing brain. Neurochemical Research. 42(6):1661-1675

Fu X, et al. 2017. Multiple Vitamin K forms exist in dairy foods. Current Developments in Nutrition 1(6). Accessed at: https://academic.oup.com/cdn/article/1/6/e000638/4558638

Fultz NE, et al. 2019. Coupled electrophysiological, hemodynamic, and cerebrospinal fluid oscillations in human sleep. Science. Accessed at: https://www.scientificamerican.com/article/deep-sleep-gives-your-brain-a-deep-clean1/

Fürst R, Zündorf I. 2014. Plant-derived anti-inflammatory compounds: hopes and disappointments regarding the translation of preclinical knowledge into clinical progress. Mediators Inflamm. Accessed at: https://www.hindawi.com/journals/mi/2014/146832/

Gardner CD, et al. 2018. Effect of low-fat vs low-carbohydrate diet on 12-month weight loss in overweight adults and the association with genotype pattern or insulin secretion: The DIETFITS randomized clinical trial. JAMA, 319(7), 667–79.

Gaytán-González A, Ocampo-Alfaro MJ, Arroniz-Rivera M, et al. 2019. Inadequate protein intake at specific meals is associated with higher risk of impaired functionality in middle to older aged Mexican adults. J Aging Res. Accessed at: https://www.hindawi.com/journals/jar/2019/6597617/

Guadagni M, Biolo G. 2009. Effects of inflammation and/or inactivity on the need for dietary protein. Curr Opin Clin Nutr Metab Care. 12: 617–622.

Gutierrez-Salmean G, Ciaraldi TP, Nogueira L, et al. Effects of (-)-epicatechin on molecular modulators of skeletal muscle growth and differentiation. J Nutr Biochem. 2014;25(1):91–94. doi:10.1016/j.jnutbio.2013.09.007

Hall KD, et al. 2012. Energy balance and its components: implications for body weight regulation," Am J Clin Nutr. 95(4): 989–994.

Hall KD, et al. 2019. Ultra-processed diets cause excess calorie intake and weight gain: A one-month inpatient randomized controlled trial of ad libitum food intake. Cell Metabolism 10.1016. Accessed at: https://www.researchgate.net/publication/333151637

Hellmuth J, Rabinovici GD, Miller BL. The rise of pseudomedicine for dementia and brain health. JAMA. 2019;321(6):543–544.

HHS & DOA (U.S. Department of Health and Human Services and U.S. Department of

Agriculture). 2015. 2015–2020 Dietary Guidelines for Americans, 8th ed.

Hidese S, et al. Effects of L-Theanine administration on stress-related symptoms and cognitive functions in healthy adults: a randomized controlled trial. Nutrients 11(10). pii: E2362.

Hoffman JR, Falvo MJ. 2004. Protein - Which is Best?. J Sports Sci Med. 3(3):118–130.

Holland BM, Roberts BM, Krieger JW, Schoenfeld, BJ. 2019. Does HMB enhance body composition in athletes? A systematic review and meta-analysis. J Strength Cond Res. Accessed at: https://journals.lww.com/nsca jscr/Abstract/publishahead/Does_HMB_Enhance_Body_Composition_in_Athletes__A.94598.aspx

Hong S, Didwania A, et al. 2009. The expanding use of third-generation aromatase inhibitors: what the general internist needs to know. J Gen Intern Med 24 Suppl 2:S383-8.

IOM (Institute of Medicine). 2006. Dietary Reference Intakes: The essential guide to nutrient requirements. Washington, D.C. The National Academies Press.

Jäkälä P, Vapaatalo H. 2010. Antihypertensive peptides from milk proteins. Pharmaceuticals (Basel) 3(1):251–272.

Johnson EJ. 2012. The effect of daily consumption of eggs on cognitive function in the elderly. Accessed at: https://clinicaltrials.gov/ct2/show/NCT01621646

Jones LC, et al. 2008. Active dehydration impairs upper and lower body anaerobic muscular power. J Strength Cond Res. 22(2):455-63.

Kalt W, et al. 2019. Recent Research on the Health Benefits of Blueberries and Their Anthocyanins. Advances in Nutrition. Accessed at: https://academic.oup.com/advances/advance-article/doi/10.1093/advances/nmz065/5536953

Katsanos CS, Kobayashi H, Sheffield-Moore M, et al. 2006. A high proportion of leucine is required for optimal stimulation of the rate of muscle protein synthesis by essential amino acids in the elderly. AJP: Endocrinology and Metabolism 291(2):E381-E387.

Kim H, Suzuki T, Saito K, Yoshida H, Kojima N, Kim M, Sudo M, Yamashiro Y, Tokimitsu I. 2013. Effects of exercise and tea catechins on muscle mass, strength and walking ability in community-dwelling elderly Japanese sarcopenic women: a randomized controlled trial. Geriatr Gerontol Int. 13(2):458-65.

Kim JA, et al. 2019. Sinensetin regulates age-related sarcopenia in cultured primary thigh and calf muscle cells. BMC Complementary and Alternative Medicine 19:287. Accessed at: https://bmccomplementmedtherapies.biomedcentral.com/track/pdf/10.1186/s12906-019-2714-2

Komaroff A. 2019. Are eggs risky for heart health? Accessed at: https://www.health.harvard.edu/heart-health/are-eggs-risky-for-heart-health

Kraft JA, et al. 2010. Impact of dehydration on a full body resistance exercise protocol. European Journal of Applied Physiology 109(2):259-267.

Kravitz L. How the body adapts to exercise. Accessed at: https://www.youtube.com/watch?v=vrOsBr64ASI

Kreider R, et al. 2010. ISSN exercise & sport nutrition review: research & recommendations. Journal of the International Society of Sports Nutrition 7:7.

Kreider RB. 2019. Essentials of exercise & sport nutrition: science to practice. Lulu Publishing Services.

LaFata D, et al. 2012. The effect of a cold beverage during an exercise session combining both strength and energy systems development training on core temperature and markers of performance. Journal of the International Society of Sports Nutrition (9)1:44.

Larsen P, et al. 2019. Evening high-intensity interval exercise does not disrupt sleep or alter energy intake despite changes in acylated ghrelin in middle-aged men. Experimental Physiology 104(6):826-836.

Layman DK, Anthony TG, Rasmussen BB, et al. 2015. Defining meal requirements for protein to optimize metabolic roles of amino acids. American Journal of Clinical Nutrition 101(6).

Longland TM, et al. 2016. Higher compared with lower dietary protein during an energy deficit combined with intense exercise promotes greater lean mass gain and fat mass loss: a randomized trial. Am J Clin Nutr. 103(3):738-46.

Macnaughton W, et al. 2016. The response of muscle protein synthesis following whole-body resistance exercise is greater following 40 g than 20 g of ingested whey protein Physiol Rep. 4(15). Accessed at: https://www.ncbi.nlm.nih.gov/pubmed/27511985/

Mansoor, N, et al. 2016. Effects of low-carbohydrate diets v. low-fat diets on body weight and cardiovascular risk factors: A meta-analysis of randomised controlled trials. British Journal of Nutrition. 115 (3), 466–79.

Martin WF, Armstrong ML, Rodrguez NR. 2005. Dietary protein intake and renal function. Nutr Metab (London) 2:25 Accessed at: https://www.ncbi.nlm.nih.gov/pmc/articles/PMC1262767/

Mate-Munoz JL, et al. 2018. Effects of ß-alanine supplementation during a 5-week strength training program: a randomized, controlled study. Journal of the International Society of Sports Nutrition. Accessed at: https://jissn.biomedcentral.com/articles/10.1186/s12970-018-0224-0

Meador BM, et al. 2015. The green tea polyphenol epigallocatechin-3-gallate (EGCG) attenuates skeletal muscle atrophy in a rat model of sarcopenia. J Frailty Aging 4(4) 209-214.

Morton RW, et al. 2017. A systematic review, meta-analysis and meta-regression of the effect of protein supplementation on resistance training-induced gains in muscle mass and strength in healthy adults. British Journal of Sports Medicine 52(6):376-384.

Mozaffarian D. 2019. Dairy foods, obesity, and metabolic health: the role of the food matrix compared with single nutrients. Advances in Nutrition (10)5:917S–923S.

Munnangi S, Angus LD. 2019. Placebo effect. Stat Pearls. Accessed at: https://www.ncbi.nlm.nih.gov/books/NBK513296/

Naghii MR, Mofid M, Asgari AR, Hedayati M, Daneshpour MS. 2010. Comparative effects of daily and weekly boron supplementation on plasma steroid hormones and proinflammatory cytokines. J Trace Elem Med Biol. 2011 Jan;25(1):54-8.

Nissen SL, Sharp RL. 2003. Effect of dietary supplements on lean mass and strength gains with resistance exercise: a meta-analysis. J Appl Physiol (1985). 94(2):651-9.

Nunes JP, et al. 2017. Creatine supplementation elicits greater muscle hypertrophy in upper than lower limbs and trunk in resistance-trained men. Nutr Health 23(4):223-229.

Omi N, Shiba H, Nishimura E, et al. 2019. Effects of enzymatically modified quercetin in supplementary protein powder on athlete body composition: a randomized, placebo-controlled, double-blind trail. Journal of the International Society of Sports Nutrition 16(39).

Onoja U, Obizoba I. 2009. Nutrient composition and organoleptic attributes of gruel based on fermented cereal, legume, tuber and root flour. Agro-Science 8(3).

Phillips SM, Tang JE, Moore DR. 2009. The role of milk- and soy-based protein in support of muscle protein synthesis and muscle protein accretion in young and elderly persons. J Am Coll Nutr. 28(4), 343-354.

Pizzorno L 2015. Nothing Boring About Boron. Integr Med (Encinitas).14(4):35–48.

Rodacki CL, Rodacki AL, Pereira G, et al. 2012. Fish-oil supplementation enhances the effects of strength training in elderly women. Am J Clin Nutr. 95(2):428-436.

Russ DW, et al. 2015. Dietary HMB and ß-alanine co-supplementation does not improve in situ muscle function in sedentary, aged male rats. Applied Physiology, Nutrition, and Metabolism, 2015, 40(12): 1294-1301.

Sale C, Saunders B, Harris RC. 2010. Effect of beta-alanine supplementation on muscle carnosine concentrations and exercise performance. Amino Acids 39(2):321-333.

Salminen A, Kaarniranta K. 2012. AMP-activated protein kinase (AMPK) controls the aging process via an integrated signaling network. Ageing Research Reviews 11(2):230-241.

Schiaffino S, et al. 2013. Mechanisms regulating skeletal muscle growth and atrophy. FEBS Journal 280:4294–4314.

Schoenfeld BJ, Aragon AA, Krieger JW. 2018. The effect of protein timing on muscle strength and hypertrophy: a meta-analysis. Journal of the International Society of Sports Nutrition. Accessed at: https://jissn.biomedcentral.com/articles/10.1186/1550-2783-10-53

Schoenfeld BJ, Aragon AA. 2018. How much protein can the body use in a single meal for muscle-building? Implications for daily protein distribution. Journal of the International Society of Sports Nutrition. Accessed at: https://jissn.biomedcentral.com/articles/10.1186/s12970-018-0215-1

Schoenfeld BJ. 2012. The use of nonsteroidal anti-inflammatory drugs for exercise-induced muscle damage: implications for skeletal muscle development. Sports Med. 42(12):1017-28.

Shai I., et al. 2008. Weight loss with a low-carbohydrate, Mediterranean, or low-fat diet. New England Journal of Medicine, 359 (20), 229–41.

Siri-Tarino, PW, et al. 2010. Saturated fat, carbohydrate, and cardiovascular disease. Am J Clin Nutr. 91(3), 502–9.

Slosman D, Vadas L, Vergnaud P,Bonjour J-P (1998) Protein supplements increase serum insulin-like growth factor-I levels and attenuate proximal femur bone loss in patients with recent hip fracture: a randomized, double-blind, placebo-controlled trial. Ann Intern Med 128:801–809.

Smith A, Gray J. 2016. Considering the benefits of egg consumption for older people at risk of sarcopenia. Br J Community Nurs. 2016 Jun;21(6):305-9. doi: 10.12968/bjcn.2016.21.6.305.

Smith GI, Atherton P, Reeds DN, et al. 2011. Omega-3 polyunsaturated fatty acids augment the muscle protein anabolic response to hyperinsulinaemia hyperaminoacidaemia in healthy young and middle-aged men and women. Clin Sci (Lond). 121(6):267-278.

Soenen S et al. 2012. Relatively high-protein or 'low-carb' energy-restricted diets for body weight loss and body weight maintenance? Physiol Behav 107:374–380.

Sophia L, Conley KE, et al. 2018. Building strength, endurance, and mobility using an astaxanthin formulation with functional training in elderly. J Cachexia, Sarcopenia and Muscle. 9(5): 826-833.

Stephen A, et al. 2014. Green tea extract attenuates muscle loss and improves muscle function during disuse, but fails to improve muscle recovery following unloading in aged rats. 118(3) 319-330.

Stohs SJ, Hartman MJ. 2015. Review of the Safety and Efficacy of Moringa oleifera. Phytother Res. 29(6):796-804.

Teixeira EM, Carvalho MR, Neves VA, Silva MA, Arantes-Pereira L. 2014. Chemical characteristics and fractionation of proteins from Moringa oleifera Lam. Leaves. Food Chem 15(147):51-54.

Timmerman KL, Dhanani S, Glynn EL, et al. 2012. A moderate acute increase in physical activity enhances nutritive flow and the muscle protein anabolic response to mixed nutrient intake in older adults. Am J Clin Nutr. 95:1403–1412.

Tipton KD, Elliott TA, Cree MG, et al. 2004. Ingestion of casein and whey proteins result in muscle anabolism after resistance exercise. Medicine & Science in Sports & Exercise 36(12):2073-2081.

Tritto AC, et al. 2019. Negligible Effects of ß-Hydroxy-ß-Methylbutyrate Free Acid and Calcium Salt on Strength and Hypertrophic Responses to Resistance Training: A Randomized, Placebo-Controlled Study. Int J Sport Nutr Exerc Metab. 29(5):505-511.

Turner C, et al. 2018. Concepts and critical perspectives for food environment research: A global framework with implications for action in low- and middle-income countries.

Global Food Security 18:93-101.

Van Horn L, et al. 2016. Recommended dietary pattern to achieve adherence to the American Heart Association/American College of Cardiology (AHA/ACC) Guidelines: A scientific statement from the American Heart Association. Circulation, 134 (22), e505–29.

van Vliet S, et al. 2017. Consumption of whole eggs promotes greater stimulation of postexercise muscle protein synthesis than consumption of isonitrogenous amounts of egg whites in young men. The American Journal of Clinical Nutrition. 106(6):1401.

Wagner M, Oehlmann J. 2011. Endocrine disruptors in bottled mineral water: Estrogenic activity in the E-Screen. The Journal of Steroid Biochemistry and Molecular Biology 127(1-2):128-135.

Wansink B, Sobal J. 2007. Mindless eating: the 200 daily food decisions we overlook. Environ Behav 39(1):106–123.

Watson E. 2018. Plant-based 'milk' row heats up as FDA commissioner says 'public health concerns' inform the agency's renewed interest in dairy standards of identity. FoodNavigator-USA.

Wilkinson DJ, Hossain T, Hill DS, et al. 2013. Effects of leucine and its metabolite ß-hydroxy-ß-methylbutyrate on human skeletal muscle protein metabolism. The Journal of Physiology 591(11):2911-2923.

Wilkinson MJ, et al. 2019. Ten-hour time-restricted eating reduces weight, blood pressure, and atherogenic lipids in patients with metabolic syndrome. Cell Metabolism. Accessed at: https://www.cell.com/cell-metabolism/fulltext/S1550-4131(19)30611-4

Wilson JM, et al. 2014. The effects of 12 weeks of beta-hydroxy-beta-methylbutyrate free acid supplementation on muscle mass, strength, and power in resistance-trained individuals: a randomized, double-blind, placebo-controlled study. Eur J Appl Physiol.114(6):1217-27.

Wu H, et al. 2015. Effect of beta-hydroxy-beta-methylbutyrate supplementation on muscle loss in older adults: a systematic review and meta-analysis. Arch Gerontol Geriatr. 61(2):168-75.

Yuan JP, Peng J, Yin K, Wang, JH. 2011. Potential health-promoting effects of astaxanthin: a high-value carotenoid mostly from microalgae. Mol Nutr Food Res. 55(1):150-165.

Zhong VW, et al. 2019. Associations of dietary cholesterol or egg consumption with incident cardiovascular disease and mortality. JAMA, 321(11):1081–95.

Zhong Y, Zeng L, Deng J, Duan Y, Li F. 2019. ß-hydroxy-ß-methylbutyrate (HMB) improves mitochondrial function in myocytes through pathways involving PPARß/§ and CDK4. Nutrition 60:217-226.

SPIRITUAL STRENGTH TRAINING

Introduction

I recognize that readers may question the inclusion of a chapter dealing with "religion" in a handbook on strength training. The purpose is to open our eyes to the idea that we possess both physical and spiritual essences, both of which can be developed. And, that we reap benefits when we spiritually strength train. The provocative question is this. We will spend a lot of time and effort training physically during our last years, but then what – we face eternity unprepared? That would not be wise, and is why we must think also about training spiritually. Please allow me to say up front that I will be sharing my thoughts regarding spiritual strength training from a Christian perspective. The reason is, Christianity is the belief system that I subscribe to. If the reader is of another faith, please realize that it is not my intention to disparage what you believe or to offend. No matter what your believe, you might enjoy reading through this chapter because the principles that I will present are, I feel, broadly applicable. Otherwise, feel free to skip this chapter altogether if you so desire – we can catch up with each other in Chapter 7.

However, I hope that many readers will stick with me, for I believe that this is the most relevant issue of all for us as older men, this matter of spiritual strength training. Although I have written this section to be read by anyone and everyone, I am thinking especially about you, the reader who may have been like I used to be – an unbeliever, and similarly for specific members of my extended family; my co-workers at Texas A&M University and those I know at other universities and from the federal government; my friends, acquaintances, and heroes in the fitness and bodybuilding arenas; my friends at the post office; the people I know who like me are fans of The Adventures of Superman; and folks who grew up watching and enjoying the sports teams and stars of the 1950s-1970s.

In this chapter, I will be drawing parallels between physical training and spiritual strength training. For example, in each, there is a goal; for each, there is a method, and for each there are adaptations that take place. In addition, on occasion I will include actual scripture verses to illustrate a point or concept. Lastly, I see it as important to also include a couple of sample prayers within this chapter. A prayer is like a conversation with God, our Creator. There is really no right or wrong way, but since some readers may not have developed what pastors refer to as a "prayer life", I wanted to assist with that.

Here is the deal in a nutshell. I am not a goodie-two-shoes who grew up in the church. As a kid, I was not pushed into attending church by my parents, or becoming a Christian. For that, I thank them – because I have seen how some of those kinds of situations turn out: the opposite of what was intended. On the other hand, I now see the value in at least pointing out certain truths to kids (and old men!) while they are being raised so that they hopefully will investigate the scriptures when they feel ready and open to receiving the wisdom they contain.

Now, I had an awareness through observation regarding the meaning of the cross, and I had seen images of praying hands and representations of Jesus on Christmas cards,

Easter cards, and post office stamps. Ultimately, I gained a perspective in my adult years that I lacked as a young person, through my own study. And that is, that the physical world and our physical bodies are only part of what exists and of who we are. There is also a spiritual dimension that should concern us, and as such, we can become interested in not only physical well-being and strength, but spiritual well-being and strength. Now as many will expect, the topic of spiritual well-being and in fact what I will refer to as **spiritual strength training**, embodies spiritual teaching from the bible.

The Ten Commandments

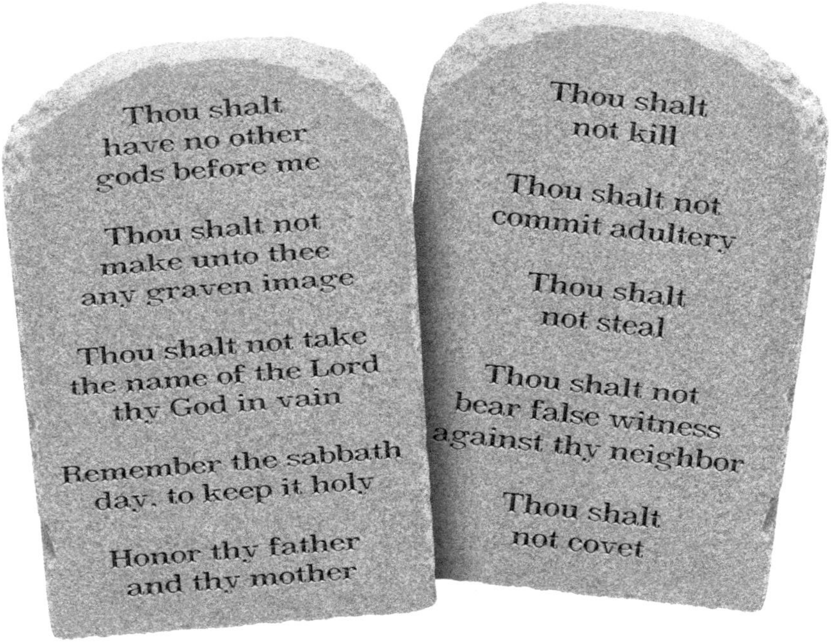

▲ **Figure 6.1** What about the Ten Commandments? (also called The Law in scripture) Notice that they either relate to our relationship with God or with our neighbor. Is it possible for anyone to keep all ten, all the time? The quick answer is no. Jesus is quoted in Matthew 5:17 as saying, "Do not think that I came to destroy the Law or the Prophets. I did not come to destroy but to fulfill". In addition, we might feel safe because we have never committed murder or adultery. Guess again! Jesus sets the record straight as follows: "You have heard that it was said to those of old, 'You shall not commit murder, and anyone who murders will be subject to judgment.' But I say to you that anyone who is angry with a brother or sister will be subject to judgment. (Matthew 5:21-22) "You have heard that it was said to those of old, 'You shall not commit adultery.' But I say to you that whoever looks at a woman to lust for her has already committed adultery with her in his heart. (Matthew 5:27-28)

For many, trying to keep the Ten Commandments represents the ultimate code of conduct, and the manner in which to live life so as to please God. They remain important in this regard, but there is more to the story. Since no human being is perfect and will stumble at some point with regard to one or more of them (this is what sin means), they actually point to the need of a special solution. James 2:10 observes : "For whoever keeps the whole law and yet stumbles at just one point is guilty of breaking all of it." And in the book of

Galatians we are introduced to the notion that the Law is actually a curse, and that it is Jesus who rescues us from this curse:

> *For as many as are of the works of the law are under the curse; for it is written, 'Cursed is everyone who does not continue in all things which are written in the book of the law, to do them.' But that no one is justified by the law in the sight of God is evident, for 'the just shall live by faith.' Yet the law is not of faith, but 'the man who does them shall live by them.' Christ has redeemed us from the curse of the law, having become a curse for us (for it is written, 'Cursed is everyone who hangs on a tree'), that the blessing of Abraham might come upon the Gentiles in Christ Jesus, that we might receive the promise of the Spirit through faith. (Galatians 3:10-14).*

My background is established in the Christian faith, so that is the perspective from which I will base my information. I of course realize that there are other faiths, and also lack of faith (unbelief), but the common denominator regarding the point of discussion is the spiritual dimension nevertheless. So the thrust of this chapter is realizing that while I can have a goal in mind when I strength train my physical body, in a similar way, I can have a goal in mind for strength training my spiritual self. In fact, the scriptures (again, my reference is the Christian scriptures, the Holy Bible) has something to say regarding the value of physical training (yes, it is mentioned) and spiritual training.

GOD'S WORD VALUES BOTH PHYSICAL AND SPIRITUAL STRENGTH TRAINING

▲ **Figure 6.2** Scripture tells us that the God who created us places value on both bodily and spiritual training. 1 Timothy 4:7-9 says, "Train yourself for godliness; for while bodily training is of some value, godliness is of value in every way, as it holds promise for the present life and also for the life to come. This saying is trustworthy and deserving of full acceptance."

Did you notice that God's word places value on physical training? That is good to hear since we are engaged in such training with the Old Man Muscle workout. But we get the added perspective from the scriptures that spiritual training (godliness) is of even more value. Both forms of training do have this in common: it takes discipline and effort to train successfully. But the difference is that the spiritual result will last forever, while the physical result is only temporary. So we really need to discuss what exactly spiritual strength training involves. How do we do it?

THE DUALITY OF EXISTENCE

Before attempting to address the specific question of spiritual strength training, allow me to talk a bit about what is called the **duality of existence**. This basically means that we have been created with both a physical nature and a spiritual nature. Further analysis of the scriptures describes body, soul, and spirit – the first of course being physical and visible while the other two are spiritual and invisible. Many writers speak of soul and spirit interchangeably, and that is fine. I will be doing the same. Also in the equation is God, the Creator, who has made us in his likeness, and yet is spirit – from which we understand, means the Holy Spirit. When we are born, the physical body and soul are united. At death, they separate, but the soul continues. This means that after we die, there is life after death for the spirit.

And the Word became flesh and dwelt among us, and we have seen his glory, glory as of the only Son from the Father, full of grace and truth. (John 1:14)

Jesus said to her, "I am the resurrection and the life. He who believes in Me, though he may die, he shall live. And whoever lives and believes in Me shall never die. Do you believe this?" (John 11:25, 26)

Now may the God of peace himself sanctify you completely, and may your whole spirit and soul and body be kept blameless at the coming of our Lord Jesus Christ. (1 Thessalonians 5:23)

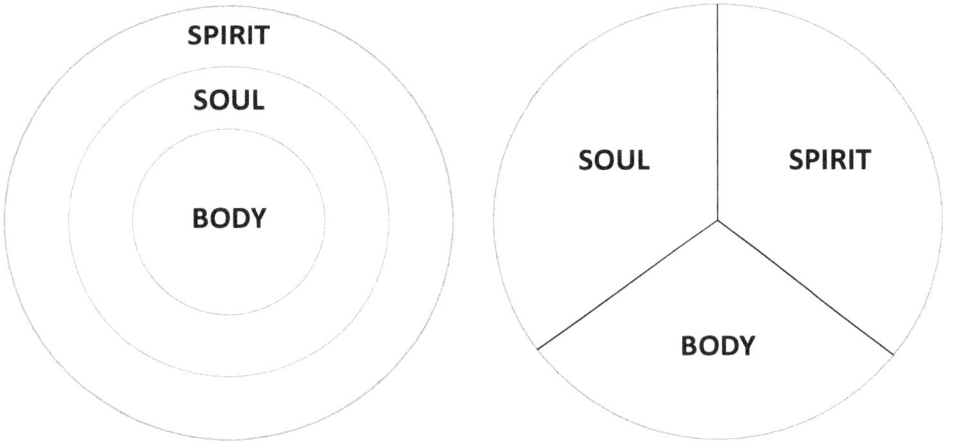

▲ **Figure 6.3** Simple conceptual representation of body, soul and spirit. The divisions are not meant to be in any particular position or proportion. The distinction between soul and spirit is typically that the soul has more to do with non-physical consciousness regarding self (one's mind, will and emotions), while the spirit is the non-physical part similar to the soul that is aware of and conscious of God. Both spirit and soul exist beyond the physical dimension, and yet all are interconnected.

Scripture asserts that the spirit, or soul, cannot be destroyed. Though the body dies, the spirit will live on. Death is not the end of conscious life; it is the separation of the body and the spirit, the second of which continues in another realm. However, there is another death that the scriptures speak of besides the physical body dying. There are two spiritual destinies: heaven, in which the spirit lives in the eternal presence of God, and hell, in which the spirit lives in eternal torment. The latter represents "spiritual death" and "the second death" because it is eternal existence in darkness, separate from God. Although I cannot be comprehensive about biblical teaching in one chapter in a small handbook like this, it bears mentioning here that not only does God have a plan for our spiritual destiny – there is also a plan for receiving a new resurrected body in eternity, one that cannot die. There is a wonderful passage on this written by the Apostle Paul to the Corinthians, that I wish to reproduce for you now:

> *The Scriptures tell us, "The first man, Adam, became a living person." But the last Adam—that is, Christ—is a life-giving Spirit. What comes first is the natural body, then the spiritual body comes later. Adam, the first man, was made from the dust of the earth, while Christ, the second man, came from heaven. Earthly people are like the earthly man, and heavenly people are like the heavenly man. Just as we are now like the earthly man, we will someday be like the heavenly man. What I am saying, dear brothers and sisters, is that our physical bodies cannot inherit the Kingdom of God. These dying bodies cannot inherit what will last forever. But let me reveal to you a wonderful secret. We will not all die, but we will all be transformed! It will happen in a moment, in the blink of an eye, when the last trumpet is blown. For when the trumpet sounds, those who have died will be raised to live forever. And we who are living will also be transformed. For our dying bodies must be transformed into bodies that will never die; our mortal bodies must be transformed into immortal bodies. Then, when our dying bodies have been transformed into bodies that will never die, this Scripture will be fulfilled: "Death is swallowed up in victory, O death, where is your victory? O death, where is your sting?" For sin is the sting that results in death, and the law gives sin its power. But thanks be to God! He gives us the victory over sin and death through our Lord Jesus Christ. (1 Corinthians 15:35-58).*

Readers who are familiar with classical music will recognize that composer G.F. Handel set several of these 1 Corinthians verses to music in his famed oratorio, "Messiah". In the section "The Trumpet Shall Sound", the trumpeter plays a solo as the bass voice sings the scripture in his aria – it is all quite moving.

Now, there is one other passage that I will share in which Jesus speaks in plain terms of the resurrection, as well as his authority in knowing the scriptures and the power of God:

> *The same day the Sadducees, who say there is no resurrection, came to Him and asked Him, saying: "Teacher, Moses said that if a man dies, having no children, his brother shall marry his wife and raise up offspring for his brother. Now there were with us seven brothers. The first died after he had married, and having no offspring, left his wife to his brother. Likewise the second also, and the third, even to the seventh. Last of all the woman died also. Therefore, in the resurrection, whose wife of the seven will she be? For they all had her." Jesus answered and said to them, "You are mistaken, not knowing the Scriptures nor the power of God. For in the resurrection they neither marry nor are given in marriage, but are like angels of God in heaven. But concerning the resurrection of the dead, have you not read what was spoken to you by God, saying, 'I am the God of Abraham, the God of Isaac, and the God of Jacob'? God is not the God of the dead, but of the living." And when the multitudes heard this, they were astonished at His teaching. (Matthew 22:23-33)*

Scripture attests to the notion that God is spirit, and yet he also became a physical body (a man) as Jesus of Nazareth, to reconcile humanity to himself on account of the separation that sin causes between creator and creation. Now, I see that we have started to get into some deep theology. Perhaps you are liking this, perhaps not so much. I will try to keep more focused on spiritual training. However, congratulations - what you just did by reading these scriptures was actually part of your spiritual strength training – reading God's word. If we read it "in the flesh" so to speak, we will have a reduced chance of understanding and appreciating it, and applying it to our lives, compared to if we read it "in the spirit". This means, we need the indwelling of God's Holy Spirit to assist us, and that can happen when we receive his spirit, in faith, through baptism. It is another supernatural process.

Peter replied, "Repent and be baptized, every one of you, in the name of Jesus Christ for the forgiveness of your sins. And you will receive the gift of the Holy Spirit." (Acts 2:38)

Recall the issue of muscle wasting that is referred to as sarcopenia. Well, it can be asserted that we suffer also from "sarcopenia of the spirit". What is the solution? Just like physical strength training can be effective against physical sarcopenia – spiritual pursuits can be effective for a weak spiritual development. Furthermore, if I can motivate myself to engage in strength training and get physical results, then I can also motivate myself to engage in spiritual strength training that produces results. But what kind of spiritual strength training?

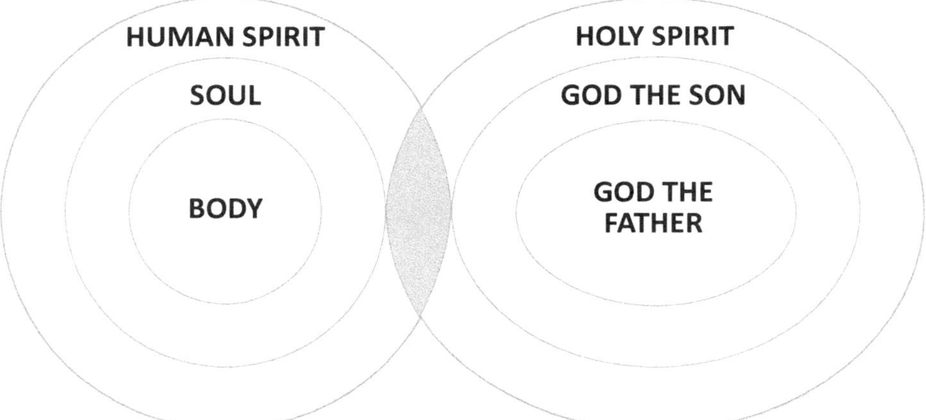

▲ **Figure 6.4** In what can only be considered a supernatural process, our human spirit can be joined by God's Holy Spirit by our invitation, in order that it receives intimate closeness and direction, to change our life according to God's better plan for it. This process is referred to as being born again. Shown is a simple conceptual representation the indwelling of God's Holy Spirit in a person, alongside his own human spirit. This is the way that God works to supernaturally transform us. Notice that just as we are a unity of three entities (body, soul, spirit), God is a unity of three (Spirit, Son, Father).

In the above figure, I mention what is now a cliché, that of being born again. On this subject, I recall many years ago as a boy seeing a commercial on television in which then football coach of the NFL's Dallas Cowboys (Tom Landry) made a statement about his personal relationship with Jesus Christ. I did not understand what he was saying, and I was confused that a person in professional sports would even talk about "religion" like that. He was a self-professed "born again believer". Many years later when I decided to first crack open a bible, I found the biblical teaching on being "born again" in the New Testament book, The Gospel of John:

Now there was a Pharisee, a man named Nicodemus who was a member of the Jewish ruling council. He came to Jesus at night and said, "Rabbi, we know that you are a teacher who has come from God. For no one could perform the signs you are doing if God were not with him." Jesus replied, "Very truly I tell you, no one can see the kingdom of God unless they are born again," "How can someone be born when they are old?" Nicodemus asked. "Surely they cannot enter a second time into their mother's womb to be born!" Jesus answered, "Very truly I tell you, no one can enter the kingdom of God unless they are born of water and the Spirit. Flesh gives birth to flesh, but the Spirit gives birth to spirit. You should not be surprised at my saying, 'You must be born again.' The wind blows wherever it pleases. You hear its sound, but you cannot tell where it comes from or where it is going. So it is with everyone born of the Spirit." (John 3:1-8)

The above scripture verses come just before perhaps the most well-known scripture in the New Testament. See if you can locate the verse that I am referring to as the conversation between Jesus and Nicodemus unfolds.

"How can this be?" Nicodemus asked. "You are Israel's teacher," said Jesus, "and do you not understand these things? Very truly I tell you, we speak of what we know, and we testify to what we have seen, but still you people do not accept our testimony. I have spoken to you of earthly things and you do not believe; how then will you believe if I speak of heavenly things? No one has ever gone into heaven except the one who came from heaven—the Son of Man. Just as Moses lifted up the snake in the wilderness, so the Son of Man must be lifted up, that everyone who believes may have eternal life in him." For God so loved the world that he gave his one and only Son, that whoever believes in him shall not perish but have eternal life. For God did not send his Son into the world to condemn the world, but to save the world through him. Whoever believes in him is not condemned, but whoever does not believe stands condemned already because they have not believed in the name of God's one and only Son. This is the verdict: Light has come into the world, but people loved darkness instead of light because their deeds were evil. Everyone who does evil hates the light, and will not come into the light for fear that their deeds will be exposed. But whoever lives by the truth comes into the light, so that it may be seen plainly that what they have done has been done in the sight of God. (John 3:10-21)

I'm sure that many readers recognized John 3:16, but it is also probably true that many did not know the context of that verse, or the further teaching that immediately followed in verses 17-21. Therefore I am happy to have had this opportunity to share them. If you are still with me, I am about to get to the nuts and bolts regarding spiritual strength training.

SPIRITUAL STRENGTH TRAINING: WHAT, WHY, HOW?

As we begin this next section, we can take note of the fact that God created our body, and that it has a purpose: to be a living sacrifice, and a place – a temple - for God's Holy Spirit to indwell.

Do you not know that you are a temple of God and that the Spirit of God dwells in you? (1 Corinthians 3:16)

But if the Spirit of Him who raised Jesus from the dead dwells in you, He who raised Christ Jesus from the dead will also give life to your mortal bodies through His Spirit who dwells in you. (Romans 8:11)

At this point we need to ask ourselves, what is spiritual strength training, why is it important, and how is it performed and achieved? Before talking about such a thing, we first must agree that we do in fact possess a spirit, and that it is responsive to inputs, and can be changed just the way the physical body is responsive to external factors. As stated just a moment ago, we have been created by God with both a physical nature and a spiritual nature in that we have a visible physical body as well as an invisible soul and spirit. Our physical bodies are born, grow and develop, but begin to decline and eventually die. On the other hand, our spiritual selves also are born, and can develop (or not), but unlike the physical, they exist eternally - that is, forever.

Furthermore, there are parallels between physical and spiritual strength training. Both forms of training are challenging if we are serious about then, requiring time, effort, and a plan. Developing life-long strength training goals requires creating and implementing a plan, as well as remaining steadfast in and continuing with the plan. The analogous situation would be that with our Old Man Muscle strength training, we develop the workout plan, we implement it two or three times each week, and we stick to it for as long as possible – or else we lose our physical strength.

Read the following and see if you identify with the complaint offered in this passage. Ask yourself if this is the way you might have felt at some time in your life.

But as for me, my feet had almost slipped; I had nearly lost my foothold. For I envied the arrogant when I saw the prosperity of the wicked. They have no struggles; their bodies are healthy and strong. They are free from common human burdens; they are not plagued by human ills. Therefore pride is their necklace; they clothe themselves with violence. From their callous hearts comes iniquity; their evil imaginations have no limits. They scoff, and speak with malice; with arrogance they threaten oppression. Their mouths lay claim to heaven, and their tongues take possession of the earth. Therefore their people turn to them and drink up waters in abundance. They say, "How would God know? Does the Most High know anything?"

The writer goes on to voice further frustration:

This is what the wicked are like—always free of care, they go on amassing wealth. Surely in vain I have kept my heart pure and have washed my hands in innocence. All day long I have been afflicted, and every morning brings new punishments. If I had spoken out like that, I would have betrayed your children.

But suddenly there is a change in tone:

When I tried to understand all this, it troubled me deeply till I entered the sanctuary of God; then I understood their final destiny. Surely you place them on slippery ground; you cast them down to ruin. How suddenly are they destroyed, completely swept away by terrors! They are like a dream when one awakes; when you arise, Lord, you will despise them as fantasies. When my heart was grieved and my spirit embittered, I was senseless and ignorant; I was a brute beast before you. Yet I am always with you; you hold me by my right hand. You guide me with your counsel, and afterward you will take me into glory. Whom have I in heaven but you? And earth has nothing I desire besides you. My flesh and my heart may fail, but God is the strength of my heart and my portion forever. Those who are far from you will perish; you destroy all who are unfaithful to you. But as for me, it is good to be near God. I have made the Sovereign LORD my refuge; I will tell of all your deeds.

Did you notice how the protagonist here speaks of the decline – physical aging - "my flesh and my heart may fail"? The very topic on our minds that we seek to address with our physical strength training. And notice too that the speaker acknowledges that he has a spirit ("my heart was grieved and my spirit embittered"). The passage goes on to speak of God as the source of strength, and we can infer that though we are not able to be near him physically, we can do so spiritually ("God is the strength of my heart… it is good to be near God").

So one major take-away from all of this, which is by the way an excerpt from Psalm 73, is that God is our source of strength. We can say that such is the case for both our physical strength and our spiritual strength.

I can do all things through Christ who strengthens me. (Philippians 4:13)

The relevance to us as readers of a strength training handbook is to appreciate that just as we can train our physical self, we can also strengthen the spiritual self. And it turns out that spiritual strength training is essential if we are to fulfil our God-given purpose in life.

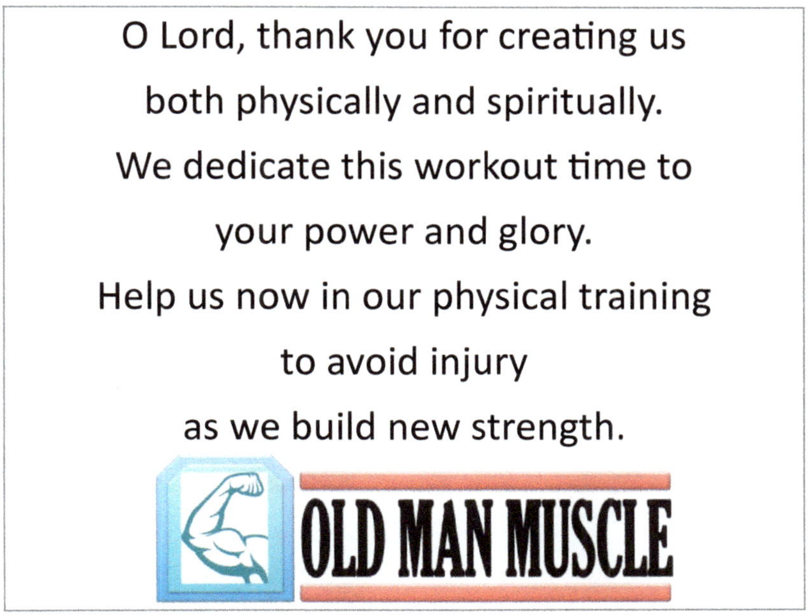

▲ **Figure 6.5** In my Old Man Muscle workout studio, I have a poster on the wall that expresses this prayer. I recite it inwardly before each and every workout, and invite anyone else to do the same if they are so inclined.

To be spiritually strong, to become "godly" and "good", we need to appreciate that goodness isn't simply the absence of "badness", because a person might do nothing wrong and yet also do nothing right, if they just sit there. Our calling is not to inaction, or to simply run away from and avoid sin and evil. What God intends and expects from us is to be active, relying on the strength he provides. One of God's gifts to us is actually to provide opportunities in life, no matter our age, to do what is right in his sight. What we do with the opportunities can become our gift back to God when we act properly. We ought to be always on the lookout for God-ordained opportunities in life – such is the true sign of spiritual maturity.

Spiritual Goals

Spiritual strength training involves specific actions that ultimately will nourish us and help us to grow in that dimension. Before we engage in these, we need to have a goal in mind – think back to the rationale for identifying S.M.A.R.T. goals in muscle training. What ought to be our goal when we embark on this journey to become spiritually strong? The answer is to focus on the glory of God and the needs of others, while becoming more Christ-like.

> *And we all, with unveiled face, beholding the glory of the Lord, are being transformed into the same image from one degree of glory to another. For this comes from the Lord who is the Spirit. (2 Corinthians 3:18)*

▲ **Figure 6.6** As I develop and mature spiritually, gaining in strength, it will be the case that the influence of Christ in me (the Holy Spirit) increases, while my natural self-centered spirit fades more and more into the background. The consequence of knowing and obeying God is that He becomes greater and greater in our life, while we become less and less. John the Baptist knew that Jesus must increase, while he must decrease (John 3:30). Even though he was a prophet of God, John knew that it was his job to lead people to the Messiah (the Christ), not to himself. So too, we are to grow to reflect more of Christ and less of our own nature as we become spiritually stronger.

To become like Jesus in character, it means not physically but in our words, actions, thoughts, and behavior. But we ought not to focus exclusively on what we are becoming. That would actually not be the correct application of scripture. And yet I have seen many so-called Christians – even pastors – focus on their own spiritual development to the point that they have missed this key teaching. And it is this – such an emphasis on self is exactly what Jesus condemned among the Hebrew teachers of the law, the Scribes, the Pharisees. His point was, in order to develop spiritually, it is your relationship to God and to other people that matters. When these relationships are done right, the person will automatically become Christ-like. How ironic – and how sad - that arrogant, self-righteous "religious" people do not understand this.

Jesus said to the crowds and to his disciples: The Pharisees and the teachers of the Law are experts in the Law of Moses. So obey everything they teach you, but don't do as they do. After all, they say one thing and do something else. They pile heavy burdens on people's shoulders and won't lift a finger to help. Everything they do is just to show off in front of others. They even make a big show of wearing Scripture verses on their foreheads and arms, and they wear big tassels for everyone to see.
(Matthew 23:1-5)

For the spiritual case, we are training in order to become more like Jesus. And by this it is meant in how we know and relate to God, in addition to how we relate to all other people on this planet. This is why Christianity is more about relationships than it is about religion. Religion embodies a series of rules and regulations that must be kept in order for a person to feel "worthy" of God. It is earning one's way to heaven. This is erroneous thinking according to the scriptures, and is what Jesus was constantly mentioning to the people of his day. In fact, although the bible teaches that faith without works is dead, the motivation for doing good is not to earn salvation, but rather is out of gratitude for the grace of God that provides salvation freely to those who are true believers and "doers" of the good works.

... those who hope in the Lord will renew their strength. They will soar on wings like eagles; they will run and not grow weary, they will walk and not be faint. (Isaiah 40:31)

For by grace you have been saved through faith, and that not of yourselves; it is the gift of God, not of works, lest anyone should boast. For we are His workmanship, created in Christ Jesus for good works, which God prepared beforehand that we should walk in them. (Ephesians 2:8)

It is obvious and logical to realize the need to understand who it is that we are trying to be like, in order to succeed. Of Jesus, Hebrews 4:15 says, "For we do not have a high priest who is unable to sympathize with our weaknesses, but we have one who has been tempted in every way, just as we are - yet was without sin." This is how Jesus has been the only one in history to keep all of the Ten Commandments, and is why he alone can serve as our "sin sacrifice".

And the Word became flesh and dwelt among us, and we have seen his glory, glory as of the only Son from the Father, full of grace and truth. (John 1:14)

JESUS IS BOTH THE GOOD SHEPHERD AND THE LAMB OF GOD

The Lord is my shepherd, I shall not want ... (Psalm 23). I am the good shepherd. I know my own and my own know me, just as the Father knows me and I know the Father ... (John 10:14-15)

'Behold the Lamb of God, who takes away the sin of the world" (John 1:29) Get rid of the old yeast, so that you may be a new unleavened batch-as you really are. For Christ, our Passover lamb, has been sacrificed. (1 Corinthians 5:7)

▲ **Figure 6.7** Jesus is portrayed in scripture as both the Good Shepherd and God's Sacrificial Lamb. In the former case, we are his sheep that have gone astray. We see the concept of a relationship rather than a religion at work. Both realities of shepherd and sacrificial lamb present eternal implications for us.

With our goal firmly in mind, we are ready to begin our spiritual strength training.

Spiritual Training Components

My list of activities that qualify as the components of spiritual strength training includes bible study, prayer, praise and worship time, communion (also known as the Lord's Supper), good works, as well as growing in faith, biblical knowledge, and spiritual wisdom. There are some others (such as participating in daily devotions) that a person might wish to add to the list. So-called devotionals are a form of private rather than congregational worship and can fit under bible study. Devotion time seeks to go beyond reading scripture only, also pondering the meaning of the verses and their specific application to life, and involves the application of prayer in response to the scripture.

> *Let the word of Christ dwell in you richly, teaching and admonishing one another in all wisdom, singing psalms and hymns and spiritual songs, with thankfulness in your hearts to God. (Colossians 3:16)*
>
> *For as the body apart from the spirit is dead, so also faith apart from works is dead. (James 2:26)*

Spiritual Adaptations

These training activities or principles, when engaged in on a regular basis over time, will lead to a training effect, and specific adaptations in the person who engages in them. Do you see how this is the identical thing that happens with physical strength training? It is scriptural that spiritual adaptations kick in when we begin with repentance. Repentance leads to the destruction of our self-centered life. When we repent, and follow through with a spirit-filled life, we live with a new focus on God and others, more than self. So now, we can identify specific outcomes associated with the training effect, what we are calling adaptations. This time, rather than being adaptations in neurons or muscle, they are spiritual adaptations. My list of spiritual adaptations includes a changed heart and transformed mind, repentance, baptism, receiving the Holy Spirit, being "born again", displaying the fruit of the spirit, and becoming spiritually strong.

> *Therefore, I urge you, brothers, in view of God's mercy, to offer your bodies as living sacrifices, holy and pleasing to God—this is your spiritual act of worship. Do not conform any longer to the pattern of this world, but be transformed by the renewing of your mind. Then you will be able to test and approve what God's will is—his good, pleasing and perfect will. (Romans 12:1-2)*

Spiritual Strength Training Results

There is one further level to examine, which follows upon the establishment of a spiritual goal, the adoption of specific spiritual strength training activities, and the causation of specific internal spiritual adaptations that take place. What I am referring to are the wholistic end results for the person who successfully trains spiritually. First and foremost, the person is assured of salvation! Salvation involves spiritual victory over sin and death, plus joyous eternal life in the presence of God in heaven. In addition, the saved person is not judged (condemned) during the event known as the Last Judgement. Rather, there is the opportunity to receive one or more crowns – or in some cases, none.

THE FIVE CROWNS OF REWARD MENTIONED IN SCRIPTURE

▲ **Figure 6.8** The scriptures compare living the Christian life to running a race (Hebrews 12:1). At the finish, there is the possibility for each one of us to receive one or more victor's crowns. I encourage each reader to look up the scripture associated with each crown, as provided in the figure, in order to better understand what characteristic of one's life is associated with each specific crown.

For everyone who has been born of God overcomes the world. And this is the victory that has overcome the world—our faith. (1 John 5:4)

Look, I am coming quickly. Hold on to what you have, so that no one will take away your crown. (Revelation 3:11)

Look, I am coming soon, bringing my reward with me to repay all people according to their deeds. I am the Alpha and the Omega, the First and the Last, the Beginning and the End. (Revelation 22:12-13)

Before we become changed, scriptures say that we are not just spiritually weak, but in fact spiritually dead. According to scripture, what follows our rebirth (being born again) is the victory that we have, and the fact that we progress from being spiritually dead to being spiritually alive. The mighty strength of the cross lies in the fact that all our sins and our spiritual death have been nailed to it, and Satan's scheme of tempting us with sin to subvert our salvation has been ridiculed, made a public spectacle of, and nullified. As I type this into the page, I want to use a gigantic font, because this point is so, so momentous and important for each and every one of us!

When you were dead in your sins and in the uncircumcision of your flesh, God made you alive with Christ. He forgave us all our sins, having canceled the charge of our legal indebtedness, which stood against us and condemned us; he has taken it away, nailing it to the cross. And having disarmed the powers and authorities, he made a public spectacle of them, triumphing over them by the cross. (Colossians 2:13-15)

I have summarized my thoughts on spiritual strength training in the figure below.

SPIRITUAL STRENGTH TRAINING

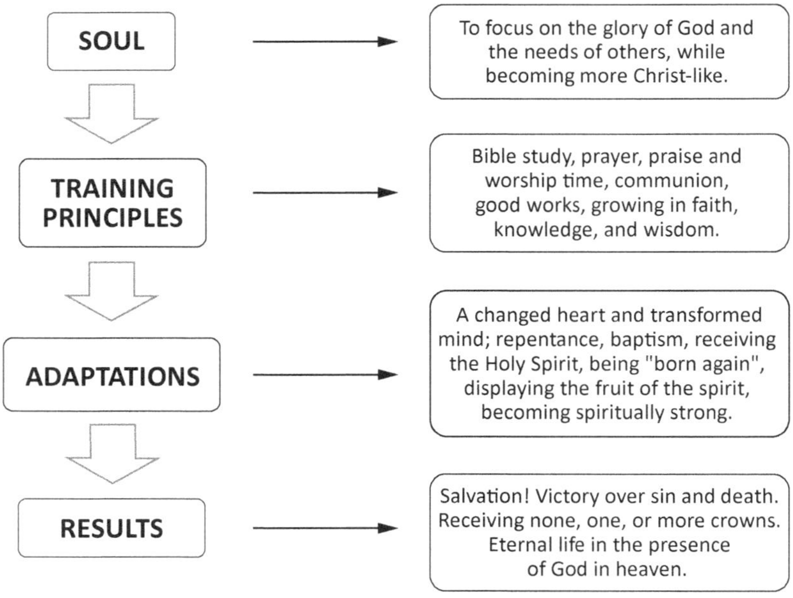

▲ **Figure 6.9** Summary chart showing the sequential organization and highlights of spiritual strength training. As with physical strength training, we can identify goals, training principles, adaptations and results when we consider spiritual strength training. In order to truly become spiritually strong, we must shift our emphasis away from ourselves and focus instead on God and other people. We perform good works not to earn our salvation, but because we have been saved by God in order that we might now do good works. It is God's grace superimposed over all of this that makes everything shown here possible.

Before we move on, I want to mention that there are tests and trials ("difficult experiences") that we experience in life that reveal our weakness or our strength. A good question to ask is, when is the best time to build strength, is it before or after the test? It would be before of course, but for many of us we get to it later. "Better late than never" is certainly true in this case.

The 23rd Psalm
The LORD is my shepherd; I shall not want. He maketh me to lie down in green pastures: he leadeth me beside the still waters. He restoreth my soul: he leadeth me in the paths of righteousness for his name's sake. Yea, though I walk through the valley of the shadow of death, I will fear no evil: for thou art with me; thy rod and thy staff they comfort me. Thou preparest a table before me in the presence of mine enemies: thou annointest my head with oil; my cup runneth over. Surely goodness and mercy shall follow me all the days of my life: and I will dwell in the house of the LORD forever.

▲ **Figure 6.10** During difficult times, it may be helpful to focus on Psalm 23:1-6.

STATISTICAL ERROR AND SPIRITUAL DECISION-MAKING

You recall that we covered type I and type II muscle fibers. Well, in this section we will examine something called type I and type II error. And in so doing, it would appear that I have generated a rather lengthy discussion. However, I felt that every word needed to be included in order to be as clear as possible on this issue. I want to make a point regarding error – such as making an erroneous decision or choice. Germane to our topic in this chapter, it refers to making an error in terms of deciding to train spiritually or not. But in order to discuss and understand statistical error, we have to first think about what a hypothesis is. A **hypothesis** is a statement about something that we believe to be true. There is also something called the null hypothesis which we must understand. The null hypothesis is often referred to as the hypothesis of "no effect" or "no difference". It is also a hypothesis that is counter to what is called the alternative hypothesis. Let's look into this a bit more before discussing Type I and Type II error.

We can use the example regarding the shape of our planet earth. Many years ago, a lot of people believed that the world was flat (we will call this the null hypothesis). However, some people began to suggest that it was in fact round (this becomes the alternative hypothesis). If we tend to believe that the world is round, our challenge is to disprove the null hypothesis, so that the alternative hypothesis "wins". This would bring us to the conclusion that earth is in fact round (a sphere), while the null hypothesis (that it is flat) is incorrect.

▲ **Figure 6.11** In statistical hypothesis testing, a type I error is the erroneous rejection of a true null hypothesis, while a type II error is the erroneous failure to reject a false null hypothesis.

The interesting part comes now. Regarding hypotheses, there are two possible cases that can turn out to be true (although they are not simultaneously ever true: it is one or the other). One is, that the alternative hypothesis is true, and the other – that the null hypothesis is true. Said another way, either the alternative hypothesis is false, or the null hypothesis is false. Again – one of them has to be true, which means that the other one is false. Let's examine this by trying a different example: a drug being used to treat a disease. Consider the following null hypothesis: "Drug A is ineffective in treating disease X". If we reject the null hypothesis in this situation, then our claim is that the drug does in fact have some effect on treating (perhaps curing) this disease.

Now, let's discuss error a bit. If the null hypothesis as stated above is found to be true based upon subsequent research and experimental data (that the drug is not effective), then we have made an ERROR in rejecting the original null hypothesis. We falsely hypothesized that drug A could successfully treat disease X, when in reality after testing, it was shown not to. Welcome to the Type I error.

This first kind of error, **Type I error**, describes when we reject a null hypothesis that is actually true. This kind of error is sometimes called an error of the first kind, or a "false positive".

Let's recap. In the example of the null hypothesis, "Drug A is ineffective in treating disease X", if we reject this statement in favor of our alternative, "Drug A is effective in treating disease X" and it turns out to be true that drug A in fact successfully treats disease X, then WE HAVE NOT COMMITTED AN ERROR.

We would only be guilty of committing an error (Type I in this case) if we reject the null hypothesis as originally stated, which subsequently turns out to be true. This also means that our alternative hypothesis must be false – because one cannot be true without the other being false, based upon statistical reasoning.

So that takes care of Type I error. What is a Type II error?

The other kind of error that is possible occurs when we do not reject a null hypothesis that is false (we should have known better!). This sort of error is called a **Type II error**, also referred to as an error of the second kind. Type II errors are equivalent to false negatives.

If we think back again to the scenario in which we are testing a drug, what would a type II error look like? A type II error would occur if we asserted drug A to be ineffective in treating disease X (in agreement with the original null hypothesis), when in reality it cured the disease. As you can see, when we agree with the original null hypothesis, there is no alternative hypothesis to be asserted.

So, what were all the possibilities again?

1. That we accept the null hypothesis as TRUE when in fact it is TRUE, which means that WE HAVE NOT MADE ANY ERROR, nor have we asserted an alternative hypothesis counter to the original null hypothesis.
2. That the null hypothesis is TRUE but we REJECT it in favor of our alternative hypothesis that turns out, is FALSE: this is a Type 1 error.
3. That we accept the null hypothesis as TRUE when it is in fact FALSE: this is a Type 2 error. We did not assert an alternative hypothesis because we accepted the original null hypothesis. Said another way: that the null hypothesis is FALSE but we ACCEPTED it.

Our next task is to apply these techniques to the notion of spiritual training, and in the broader context, to the question as to whether the scriptures are true in what they say or not. We do this because if we are agreeing that spiritual training has value, then we are agreeing with the Bible. If we are agreeing with the Bible, we must accept all of it, not just bits and pieces – because the Bible makes its own claim to be the Word of God. If the Bible turns out to be true, then faith is important, and there are eternal implications for each and every one of us based upon our faith or lack of faith. So now, we will consider the various possibilities, calling them "cases".

CASE 1: Null hypothesis is: "The scriptures (Bible) is the truth" when in fact this is TRUE, and we say it is TRUE (no alternative hypothesis): NO ERROR "we have faith that proves valuable"

CASE 2: Null hypothesis is: "The scriptures (Bible) is the truth" when in fact this is TRUE, and we say it is FALSE (alternative hypothesis: you can believe anything you want): TYPE 1 ERROR "we are without faith"

CASE 3: Null hypothesis is: "The scriptures (Bible) is the truth", when in fact this is FALSE, and we say it is TRUE: TYPE 2 ERROR, "faith that is in vain"

CASE 4: Null hypothesis is: "The scriptures (Bible) is the truth", when in fact this is FALSE, and we say it is FALSE (no alternative hypothesis): "we are without faith"

Notice that there are four possible outcomes in terms of what we choose to believe. We end up with either (1) valuable faith, (2) no faith, (3) faith that is in vain, or (4) no faith (again). So, this pans out as one positive outcome choice and three that are either neutral or negative – depending upon what reality turns out to be. The big question is, with so much potentially riding on this issue, is it smart to make an error of any kind regarding whether or not to believe in the scriptures?

What do I mean by "potentially riding on this issue"? Well, it comes down to either (a) the scriptures are true, or (b) the scriptures are not true. If the scriptures are not true and we don't place our faith in them, then we neither gain nor lose anything (CASE 4). If instead the scriptures are not true but we choose to believe in them, we are believing a fairytale, and again, we gain nothing but lose all the time and energy put into our faith (CASE 3). On the other hand, if the scriptures are true, then here is what is hanging in the balance ("riding…"): all that is said regarding heaven and hell, sin, salvation, Jesus as Lord and Savior, the resurrection and eternal life – they all must be true. This is because the scriptures declare that every word they contain (as they come from the Creator) are true, which means we cannot pick and choose which parts to believe and which not to – it is all or nothing.

> *All Scripture is breathed out by God and profitable for teaching, for reproof, for correction, and for training in righteousness, that the man of God may be competent, equipped for every good work. (2 Timothy 3:16-17)*

> *The words of the LORD are pure words, like silver refined in a furnace on the ground, purified seven times. (Psalm 12:6)*

> *Every word of God proves true; he is a shield to those who take refuge in him. (Proverbs 30:5)*

> *For the law was given through Moses; grace and truth came through Jesus Christ. (John 1:17)*

So, we are left with either we accept the truth and allow our lives to be enriched by it (CASE 1), or we reject the truth and endure the consequences of that mistake (CASE 2). Put plainly, this little exercise shows us how easy it is to commit either type of error in anything – including our spiritual decision-making. And that brings us down to the realization that the real choice left is between believing in the information presented in the scriptures, or not. What makes it all the more urgent and critical a decision is the possibility that the null hypothesis "The scriptures (Bible) is the truth" turns out to be true. Then, if we reject this hypothesis and fail to cultivate a spiritual life of faith, we will not participate in all the joys and promises as detailed in the scriptures.

As I prepare to close this section, we can consider that time as we know it includes events of the past, the present, and the future. Concerning the past and Jesus of Nazareth – who was he, what did he say, and what did he do? Concerning the present and Jesus of Nazareth – where is he? What does he want? And concerning the future and Jesus of Nazareth – what is he going to do? Thinking back to committing errors, we might think about these questions, and some others: What are my options, spiritually speaking? What do I risk if I do not accept that the scriptures are true? What do I gain if I accept that the scriptures are true? What specifically should I do? My brothers, I want all of us to make up our minds on such key issues with our eyes wide open, realizing what is at stake, and taking the appropriate next steps.

One further observation on the topic of error. You noticed that we used the word "reject" in our discussion of committing Type I or II errors. It is highly important to be aware that the Living God and Creator of all things has provided us with free will. This enables us to either accept or reject Him at His word. Which means, God has created us in such a unique way that he has given us the power to reject His gracious and merciful love – the most amazing and awesome commodity in all of creation - if we so desire. This is simply mind-blowing when you consider what it means - and truly tragic, when "I reject" is the decision that any individual makes.

DYING TO SELF

I watched a video available at rightnowmedia.org recently and it made the following point: probably the most difficult choice to make in life is choosing to follow the Lord (Jesus, the Christ) instead of following self. Here's the difference. Followers of self are too busy, too distracted, and too slow to listen to and respond to the Lord – or to even be interested. Instead, followers of oneself listen to the world around them. And they continually need down time and fill such times with hobbies, pleasure and so forth. On the other hand, followers of the Lord are filled by his teachings. They respond swiftly because they know the difference between needs and wants. Sure – they can still enjoy hobbies and pursue interests and such but the difference is, followers of the Lord prioritize differently. They put others before self, and the Lord above all. Such people choose to pick up their cross and follow him daily. The scriptures put it this way (Matthew 16: 24-26):

> *Then Jesus told his disciples, "If anyone would come after me, let him deny himself and take up his cross and follow me. For whoever would save his life will lose it, but whoever loses his life for my sake will find it. For what will it profit a man if he gains the whole world and forfeits his soul? Or what shall a man give in return for his soul?*

▲ **Figure 6.12** The hierarchy of serving: Lord > others > self. In addition, realize that we serve God by serving others: "You, my brothers, were called to be free. But do not use your freedom to indulge the sinful nature; rather, serve one another in love." (Galatians 5:13)

To take up our cross means to follow in the path of righteousness that Jesus showed. His cross led him to Calvary, where he atoned for the sins of the world, even at the cost of his own life. For us, to deny oneself and take up one's cross does not mean to check out of society and become a hermit, nor to anticipate and endure personal hardships and tragedy on a daily basis. It means to "die to self". It means to be on a life mission to put others above ourselves, and God above everything, regardless of the cost to our personal lives (even to the point of losing one's life, if necessary).

We also see from Matthew 16 verse 26 what is at stake: it is our very souls. Notice that it is we who must take up our own cross - no one else can do it for us, nor can we do it for someone else. We each are individually responsible on this account. "**Dying to self**" may sound like a cliché, but that is what taking up our cross is all about. It is one of the most challenging things to do, and toughest decisions to make, here on planet earth. And make no mistake – it is a daily decision. But the consequences are all positive regarding the difference made in people's lives here and now, as well as for all eternity. Not because I say so, but because the scriptures say so. And here's the amazing part - God provides us with his Holy Spirit and his Holy Word to help us achieve this goal, if we would only allow such things to happen in our self-centered lives.

> *"Everyone I know who has been greatly used by God has gone through an experience of 'dying to self' as described in scripture. (Pastor Bill Bright, founder of "Campus Crusade for Christ). Related scripture: I have been crucified with Christ and I no longer live, but Christ lives in me. (Galatians 2:20).*

Somewhat paradoxically, it is not until we know the reality of "death to self" that we can begin to live for Christ, allowing God to truly use and bless us. Being spiritually strong requires a daily decision to surrender to the lordship of Jesus, the Christ – due to all of the distractions and evil in the world and the daily temptations that surround us. It involves yielding our will to God and examining the life of Jesus, who is the visible expression of God's holiness for us. True life – abundant life – begins with dying to self, which is a liberating action that produces the joy and peace that former LSU and NBA legend Pete Maravich spoke of. Pete, as we will read in the next section, made the decision in time, and became spiritually strong, which is most important both on this earth now, and in terms of eternity. The question is, are we each ready to work toward the same goal?

THE EXAMPLE OF PISTOL PETE MARAVICH

I'd like to close out this chapter by talking a bit about one of my all-time basketball heroes, "Pistol Pete" Maravich, who attended LSU in the late 1960s and played in the NBA for the Hawks, the Jazz, and in his last season in 1980, for the Celtics (with rookie Larry Bird). Not many people today would know that it was Pete (not Wilt Chamberlain, not Michael Jordan, not Magic Johnson, etc.) who set the all-time college basketball scoring record by averaging 44 points per game for all three of his varsity seasons – an unbelievable record that stands even now (in 2020) and indeed may never be broken. He also has the distinction of becoming the first pro athlete in any sport – be it baseball, football, basketball, hockey, you name it – to be given a million dollar contract. The very first. And he is the youngest player ever to be inducted into the basketball hall of fame. But these achievements are not all that make Pete Maravich outstanding. Toward the end of his brief life, he became a man of spiritual strength, and serves as a role model for us all even today, as a result.

After a life of incredible achievement as well as incredible pressure, disillusionment, and frustration, Pistol Pete changed and made a very public Christian testimony, and you can watch it on YouTube. I highly recommend that you take the time to view it (the video is under ten minutes). Pete, who never considered himself eloquent or articulate, speaks very well and without notes. At one point he touchingly recalls his own father, that "he wasn't a spiritual giant" but that he "knew who Jesus is" and that when he died, he had peace, with Pete by his side. He said that God used his dad to teach him about faith.

Pete has recounted that when he was playing in the NBA, he was an alcoholic and struggled with many personal and relational issues, and that he mocked Christians and Christianity. He went on to say how he had become bitter and despondent after he retired from the game, and that he had hatred for some people. He tried to find meaning to life in all kinds of things, from belief in UFOs to becoming a vegetarian, and also via eastern mysticism. He was searching for meaning, and especially for peace, and the joyful life that the world says that money and fame will buy, but that he realized was eluding him. Eventually, he reached a point at which he became suicidal, and was about to take his own life, when one evening he heard the voice of God telling him to "BE STRONG" and to lift his own heart. It is a compelling testimony, because Pete immediately sat up and took notice, and decided to change his life right then and there. He speaks about the joy he received that he had never previously experienced, not with all his money, his fame, his trophies, his collegiate scoring record, his Hall of Fame ring, none of it. He said he would gladly trade it all for knowing the Lord Jesus and the joy only He can fill your heart with. And he did! Pete Maravich had finally found the joy and peace that had been missing from his worldly life.

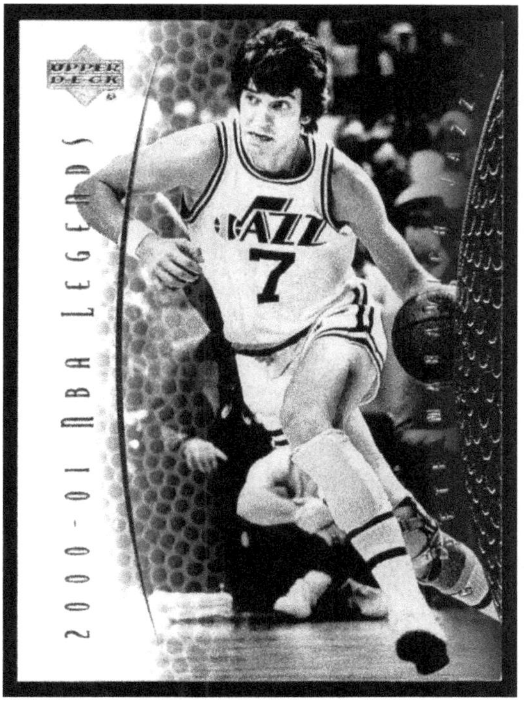

▲ **Figure 6.13** I own several collectible trading cards of vintage NBA stars, including these of Pistol Pete Maravich.

And Pete went on, in the few short years that remained for him on this earth (he died at age forty during a break in a pick-up game with Dr. James Dobson, a well-known Christian pastor, on January 5, 1988), to witness to others and share his testimony. He was able to touch many young lives at his summer basketball camp. Also he reportedly witnessed to N.Y. Yankee great Mickey Mantle, who at the time was dying of liver disease. It was documented by the doctors after Pete's own death that he was born with a rare heart condition that should have ended his life around the time he was playing his last college games or certainly by the beginning of his pro career. This is especially compelling considering the intensity with which he played the game of basketball and the abnormality of his heart. But God had plans for Pete Maravich – as he does for you and for me – and he gave Pete enough time and the chance to change his life around, which Pete did, to his eternal credit. And most importantly, he had enough time to share his testimony and serve as an example to others. He used his remaining time so very wisely – which unfortunately so many do not. And notice that Pete had an undiagnosed condition – and died suddenly and without warning. The good thing is, he prepared himself spiritually, and was "ready". Can you and I say the same?

> *For I know the plans I have for you declares the Lord, plans to prosper you and not to harm you, plans to give you hope and a future. (Jeremiah 29:11)*

What an inspiration his story is! Far from being the tragic end of an athlete dying at too young an age, it is a triumphant story of a man standing at the door of death, who was given a second chance at life – and that he made the right decision by taking God at His word. At the retirement ceremony of his Atlanta Hawks jersey in 2017 (he had worn number 44), which incredibly took place nearly thirty years after his death, his wife Jackie told the crowd that in spite of all his college and pro basketball successes, "Pete met his ultimate goal when he became a Christian." Notice, she did not say that Pete met his ultimate goal when he scored 68 points in one NBA game. Instead, he proved himself to be spiritually strong, because the measure of spiritual strength is in changing one's life toward knowing the God of the bible and serving others. The way that scripture tells us this is demonstrated is through what is called "the fruit of the spirit." Pete's life, and his personal character, had changed to the point that he was achieving the likeness of Jesus and living the fruit of the spirit in his daily life. He had become a spiritual strongman.

So as I close this chapter on spiritual strength training, I leave you with the example of Pete Maravich. The change that came over him is possible for each one of us. Only a select few individuals have natural, superior athletic ability of the kind that Pete Maravich displayed in order to make it in pro sports. But Pete Maravich represents even more of a rare person in that he not only reached the pinnacle of athletic ability in his sport, but also in his spiritual life, which will serve him for all eternity. The former required giftedness coupled to years and years of hard work, while the latter required only a few seconds of time to realize what was at stake, and to make the decision that led him down the path to become "a new creation". To accept God at his word, and to live a changed life.

What we need is the same desire to make this our goal. We do not need physical and athletic giftedness in order to become spiritually strong. We need to accept God's holy word as the truth, and allow it to change us as we learn and apply it in our daily life. If you are at the point in your life where you realize that you need to make this change, here is a simple prayer that you can use to start the process. This should be followed up by joining a local group of believers – a church – in order to continue to grow spiritually, to assemble with others to worship and sing praises to God, and in order to live the kind of life that God has planned for you.

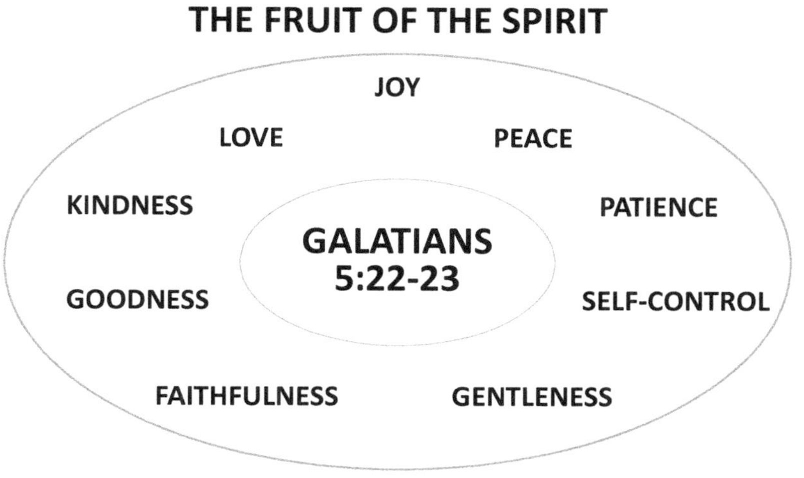

▲ **Figure 6.14** The fruit of the spirit. It was pointed out to me years ago that certain individuals who do not have the Holy Spirit in their life can exhibit one or several of these qualities, but only with the Holy Spirit is it possible to achieve all nine.

"O God in heaven, I know that it is sin that separates me from You. You are holy and righteous, but also abounding in grace and love. You sent your Son Jesus to die on the cross as the penalty for all human sin, including mine. And through Your power, Jesus resurrected from the dead, conquering both sin and death. So now I want to be a part of your family. You say in Your Word that if I acknowledge that You raised Jesus from the dead, and that I accept Him as my Lord and Savior, I would be saved. So God, I now say that I believe that Jesus died for my sins, that You raised Jesus from the dead, and that He is alive and I want to become more like Him. I accept Him now as my personal Lord and Savior. I accept my salvation from sin right now. I will be obedient and will be baptized as an outward sign of this decision, and to receive the gift of the Holy Spirit so that I can lead a more Spirit-directed life of service. Thank you, Father God, for forgiving me, saving me, and giving me eternal life with You. In the name of Jesus I pray, Amen"

Did you know that one of the joys that we have when we become spiritually strong is to actively pray regularly and persistently on behalf of others, focusing on them and their needs? Philippians 4:6 tells us "Do not be anxious about anything, but in everything by prayer and supplication with thanksgiving let your requests be made known to God." If we routinely pray, we become less self-centered and begin to see opportunities in life to serve others rather than ourselves, as we serve the Creator, who wants us to live a godly and spiritually strong and productive life. Changing our behavior in this manner actually becomes an act of worship. And we also thank God each and every day for all the blessings of life. Though we suffer and endure hardships in life, what God has provided through his Son Jesus is beyond comparison, and makes all the pain of this present, fleeting life endurable.

A PRAYER FOR HEALING
Dear Heavenly Father, I lift up my brother John Doe in prayer. As you know, John is facing a health challenge of some seriousness, and my prayer is that you please provide for complete healing of his physical condition, whether by your Almighty power, or through the efforts of the medical professionals caring for him. O Holy Spirit, come and rest upon John like a dove, and protect and shield him during any treatments that he may receive, and throughout his recovery. Almighty God, you are the water of life; you are a fresh spring; you are a healing rain to those in need. Your word says in Jeremiah, "Heal me Lord and I will be healed. Save me and I will be saved. For you are the One I praise." Thank you in advance for the restoration that only you can provide. We rest beneath your mighty wings of love, healing, and power, and praise you for your eternal goodness and mercy. In the name of Jesus I pray, Amen.

▲ **Figure 6.15** One of the manifestations of having been changed by God is having compassion for others. This includes actions such as interceding on their behalf through prayer. Shown is a prayer that I use daily, and want to share, so that you can adapt for your own personal use.

▲ **Figure 6.16** The Old Man Muscle fitness studio displays several large photos of Steve Reeves as Hercules, coupled to inspirational scripture (e.g." Be strong in the Lord, and in the power of his might." Ephesians 6:10), and in this case, an additional decorative touch - the heavy chains.

RECAP AND WHAT'S AHEAD

Just as we have been called to get onboard with the Old Man Muscle strength training program to transform us physically, I suggest that we have also been called to get onboard with a spiritual strength training program, one that has as its purpose to ultimately transform us into the likeness of Jesus, in character and purpose. In this chapter, we have examined the value and components of spiritual strength training. When coupled to physical strength training, the combined result for the person exceeds their simple sum. Furthermore, I believe that a personal trainer really misses an essential opportunity unless he provides training for the whole person: both body and spirit. This is because we are created to be in spiritual balance with our physical selves. Muscle training is very important in this life, but will only get us so far when we consider eternity.

We also can appreciate that the things of this world should be of little value to us. As Pete Maravich learned and attested to, the accumulation of wealth, career advancement, and earthly pleasures have no lasting value. Only a life which gives glory to God has lasting value. Rearrange the priorities of your life, and look forward to Christ's coming with great confidence. Live your life in accordance with the example set by Him, and you will experience great reward for all of eternity. The time to do so is now, for we never know when our window of opportunity will close forever.

The next chapter is our final one. The title of it, "A Better Life at Last", can ring true for any of us willing to make the move to physical as well as spiritual strength training. While some may live their final days full of **regrets**, this is avoidable; we can live without regrets. In the last chapter, I will not so much summarize the suggestions laid out in this handbook, as I will examine some thought-provoking topics related to being a man of older age. I hope that you will find them illuminating and of value.

RESOURCES

Farrell K. Nourishing Devotions: Strengthening our Spiritual Muscles. Accessed at: https://dashingdish.com/video/nourishing-devotions-strengthening-our-spiritual-muscles-to-overcome-temptation

Illowsky B, Dean S. 2013. Introductory Statistics. Accessed at: https://openstax.org/details/books/introductory-statistics

Stedman R, Authentic Christianity. Accessed at: https://www.raystedman.org/thematic-studies/body-life/the-goal-is-maturity

Maravich P. YouTube testimony. Accessed at: https://www.youtube.com/watch?v=Y5s6xIZfQtQ

Scripture verses (in general) Accessed at: https://www.biblegateway.com/passage

Stott JRW. 2006. Basic Christianity. IVP Books.

Warren R. 2002. The purpose-driven life. Zondervan Press.

A BETTER LIFE AT LAST

Introduction

Welcome to the last chapter in the Old Man Muscle handbook. We have covered a lot of ground within these pages, and I trust that you will come away with a desire to implement some of the strategies that I have shared. If you are in your 50s, it is quite likely that you have 30 or more years of life ahead of you. During that span of time, there no doubt will be breakthroughs that will impact the manner in which people age, and there will be new information on exercise and nutrition to help seniors remain vital and functional in spite of advanced years. If you, like me, are older than 60, what we have to go on for the present is highlighted in this book. We can choose to act on the information, and see where it takes us. We may be in for a pleasant surprise, compared to the alternative of just allowing things to happen.

What I will cover in this relatively brief chapter are various topics that when taken and applied should have each of us aging more gracefully in a way that allows us to do more than simply exist, but to live to our maximum longevity, feeling satisfied, finding meaning, and being happy. This handbook obviously places an emphasis on strength training. I do realize however that not everyone, for one reason or another, will be able to strength train either physically or spiritually in the manner described. That does not mean that such a person "fails" or is to be somehow ashamed or feel like a "loser". We all have differing abilities, opportunities, and motivations that apply to how we make decisions and ultimately live our lives. In this handbook, I have offered strategies that can be tried if one is interested in healthy aging - some might call it "successful aging". Also, we ought to realize that the physical self is temporal, and that who we are aside from the physical is what becomes eternal, and is most important.

HELLO FELLOW BABY BOOMER… or is it "OK BOOMER!"

Baby boomers like me were born during the baby boom era in the years following World War II (1946-1964). We were born into the exciting post-war era of space travel, Swanson tv dinners, cigarette commercials, Nat King Cole, Elvis, The Adventures of Superman, and hot rods.

GEN Z	**Millenial**	**GEN X**	**BOOMER**
Born 1995-2015	Born 1980-1994	Born 1965-1979	Born 1946-1964

▲ **Figure 7.1** Comparison of birth year ranges between Gen Z, Millenial, Gen X, and Baby Boomer.

We boomers have been the dominant generation for years, but the tide has already turned: millennials (73 million) outnumbered boomers (72 million) as the most populous USA generation as of 2019. However, it has been asserted that Gen-Z is already surpassing millennials. Interestingly, Gen Z has been suggested to share the same generational personality as the hardworking, patriotic, sensible, loyal, Silent Generation, which was offspring to the Greatest Generation. The Greatest Generation was the first generation born in the 20th century, and lived through the Great Depression and fought in World War I.

So, years ago we boomers were the new kid on the block, and became the biggest - but as time rolls on, each generation ages and the next ones move into positions previously held by the earlier generations. And that brings me to "**OK Boomer**". Had you heard of it? What I know of it is, the younger generation, which in this case means Millennial, apparently blames the present day ills of our world (wars, shady politics, global warming and other environmental difficulties) on the boomers. This is because we have had boomers leading the nation for the last bunch of presidential terms. It has led to the mocking, "OK Boomer" response to the baby boomer generation. Boomers are viewed as old and out-of-touch condescending authority figures, who muck up the planet and give advice that has little relevance to the younger generations. Makes me think of the old tv show "Father Knows Best", except, in reverse.

The flip side of this generational conflict has boomers viewing millennials as clueless, disrespectful, gadget-clutching and financially irresponsible toddlers. Experts suggest that what "OK Boomer" is really about involves economic anxiety on the part of millennials, a generation resistant to change (boomers), and superimposed upon these traits, the threat of environmental collapse.

▲ **Figure 7.2** The usage of the phrase "OK Boomer" serves to highlight a generational divide. It has come to indicate an older person who is close-minded and resistant to change.

This OK Boomer thing can be considered as an example of ageism. **Ageism** is stereotyping and discrimination based upon a person's age. And it would appear to operate in either direction, meaning that a younger person as well as an older person may each experience it.

What both sides really need to know and appreciate is that, as with many stereotypes, what these generations are said to think of each other is not totally accurate in real life. Many boomers appreciate and value what the millennials of their acquaintance are all about. Rather than being slovenly and clueless, many millennials have proven to be outstanding citizens with ample ability, motivation, and skills. Similarly, not all millennials dismiss baby boomers. Millennials often seek out advice from boomers, and base their decision-making in part on the wisdom their elders provide. World views may still differ, but these are significantly shaped by our generational experiences and biases. The old saying of "agreeing to disagree" on certain things while still being not only tolerant of one another but willing to co-exist and learn from each other, applies.

The previous paragraph hinted at the benefits of age and experience. We have all probably heard the saying, "with age comes wisdom." While not universally true, most people who are older have chalked up more life experiences, including successes and failures, compared to a younger person. With the longer life comes more information to draw upon, plus the ability to recognize patterns (regarding behavior patterns of other people, the manner in which situations typically unfold, and so forth). The more patterns one can recognize, the better chance at being "wise" at decision making. However, the age connection is disputable, because there are older people who make terrible choices in life and do not appear to learn lessons, while at the same time there are some young folks who are amazingly capable and live wise lives beyond what you would expect considering their age.

HOW TO BE HAPPY AND HAVE NO REGRETS

It may sound simplistic, but in order to be happy, being a positive person is what we ought to strive for. Not only does that make us attractive to other people – no matter their age or ours, but it helps us function better on a daily basis. I'll never forget the impact that our neighbor, former Texas A&M University and professional baseball player Wally Moon had on me. My wife and I moved into our present residence around 2014, and soon discovered that Mr. Moon was our neighbor. Well, he was a never-ending source of inspiration to me due to his sage advice and sunny demeanor. And he lit up when I mentioned to him that I had collected his L.A. Dodger and St. Louis Cardinal baseball cards in my youth. I even had the opportunity to tell him about my plan to become a fitness trainer of older men one day in the future. In his eighties at the time and occasionally using an oxygen tank, he immediately said, "I'm interested – sign me up!" Now that is a person with a positive and happy outlook on life. God bless you, Wally!

The flip side of being positive is of course negativity. It has been said that life is too short to be around negative people. These people can be mad at the world, and have deep regrets regarding how their life turned out. They complain all the time and play the victim. Chronic negativity in a person drags everyone around that person down, and if this is a family member, it can make family gatherings quite unpleasant. Rather than spiraling down the negativity slide ourselves, even when tough times come, we can still appreciate what we have, and try to make the most of everything. If things are so extreme that we are at the end, then we certainly can focus on the blessings that are in store "on the other side" - no pain, no disease, no death – just wonder, awe and fellowship. We should make it a point not to be that negative person – one less bit of negativity in this world would be a good thing.

POSITIVE AGING'S STATE OF MIND

- Appreciate and approach each day with a sense of purpose
- Adapt and cope with life's changes
- Maintain a positive state of mind
- Be optimistic; avoid negativity
- Be courageous

▲ **Figure 7.3** Our state of mind will drive our feelings toward being happy or not – even when we are old and frail.

ON MAKING ADJUSTMENTS AND RESILIENCE

Making Adjustments

Why do we have to make adjustments? Because with aging come a host of changes. Some people experience them sooner while others do so later. But come, they will. These include changes to our physical bodies – even if we eat healthfully. Our vision, hearing, and memory decline. We may become incontinent. We have balance and mobility problems. Our body strength and stamina are not what they were at age twenty - which is why we strength train. In addition, our bones and joints may ache and deteriorate. As one physician reportedly told his seventy-year old patient, "All my patients your age who are free of pains are dead!" We might add in the vernacular, that reality bites.

CREATIVE AGING

- See possibility instead of problems
- See aging as a catalyst for rich new experiences
- Renew passions and reinvent yourself

▲ **Figure 7.4** Have you ever heard of creative aging? In a nutshell, it involves critically thinking about age-imposed limitations, and coming up with "possibility strategies" – behavior modifications – that allow for continued functioning, even if in different ways.

In a book by Dr. Marc Agronin entitled, "The End of Old Age", there is the account of concert pianist Arthur Rubinstein, who performed up until his 89th birthday. Reportedly, Rubinstein "dealt with age-induced declines in his skills by selecting a more limited repertoire, optimizing his performance through extra practice, and compensating by altering his tempo during certain sections to highlight the dynamics of a piece." This approach is what some have termed "creative aging", and is something we can all attempt to some degree.

Resilience

In addition to the need to adjust, there is this truism to consider: "Adversity doesn't discriminate". All of us will have to deal with tough times. Depending upon our age and life circumstances, this may have happened already – multiple times. Any and all of us may experience moments in life when the path we thought we were taking makes a change of direction in ways we neither want nor ever imagined, sometimes with painful or even tragic results. These are the tough times in life. In spite of it all, there is this thing called **resilience**. It is the ability to regroup and get back on track with our lives in spite of adversity.

Being resilient starts with the realization that struggle is part of life. How to react and respond? Experts advise that we are not to allow our attention to become fixated exclusively on the negative that is happening or has happened. Also, we are to consider if the way we are thinking and acting in the present is helping or hurting us. If the latter, we must not allow that. Though not an easy behavior to implement, behaving this way can help us rebound from setbacks. It also should be abundantly clear from the previous chapter that the role of faith and spiritual strength in facing adversity is undeniable.

THREE ELEMENTS ESSENTIAL TO RESILIENCE
• View challenges and setbacks as learning and growth opportunities rather than things that destroy you. • Be committed to things that matter most to you - relationships, strength training, spiritual growth - and those setbacks will take a back seat in your life. • Focus on what you have control over- don't worry about things beyond your control.
CHARACTERISTICS OF RESILIENT PEOPLE
• Resilient people see the effects of negative life events as temporary rather than permanent. • Resilient people do not allow negative life events to affect or displace other unrelated areas of their lives where they are successful and thriving. • Resilient people do not blame themselves when negative life events occur; they perform damage control, regroup, and move forward - even if slowly.

▲ **Table 7.1** Elemental components of resilience and the attributes common to those who are resilient.

CONSIDER REINVENTING YOURSELF, AND JOURNALING

Reinvention

The NIA (National Institute on Aging) has issued a report, "Growing Older in America: The Health and Retirement Study". The report says that over 70% of Baby Boomers want to work during their retirement. This often is achieved via a gradual transition from full time to part time employment, and eventually to full retirement. This is in sharp contrast to transitioning from full-time work to full-time retirement. The point is, what the Boomers who go into "semi-retirement" must consider is, what work will I be performing? This gets at the notion of reinventing oneself.

Reinvention is the process in which we recognize the changes happening to us and around us, which cause us to think critically about the direction of our life, and the need to adapt to the changes. Through the process of reinvention, an older man might transition from his full-time career to a completely different one that is part-time. Depending upon when you exit full-time employment, this reinvention process would kick in perhaps at age 60, age 65, or even age 70. Many experts feel that it is not only a possibility, it is a financially sound strategy that is also a healthy strategy for the person at retirement age. Due to what is called the **paradox of aging** – that is, the older we get, the happier we become – you may find that your retirement career is much less stressful and is more enjoyable than its predecessor.

Reinvention relaunches our waking and working hours in ways that tend to make us smile. We spoke of actor Clint Eastwood earlier in this handbook. Well, he is a good example here. As he aged, you will notice that he did more directing and less acting. He still had plenty of ideas and scripts to deliver to audiences – but it was from the opposite side of the camera. It is almost as if his acting career was training for, and led up to, his directing career. Mr. Eastwood had successfully reinvented himself, and in his case, within the same profession. He has stayed active "in the Hollywood game" well into his eighties, but in a different way.

Journaling

Now on to the topic of journaling – what, why, how, when? **Journaling** is the process of writing down thoughts, experiences, observations – typically into an actual journal (notebook) specifically reserved for this purpose, although nowadays digital media are often used. Digital journaling may sound a lot like blogging, and it is similar. Although journals may tend to be reserved for more private matters and not fully shared on the internet, which makes them more like a diary, blogs are musings specifically written for public consumption. The benefits of journaling include: what you write about yourself becomes "permanent" and is similar to an autobiography; your siblings (brothers and sisters) will want to read it to see if they are ever mentioned, and to relive certain experiences through your recollection of them. In addition, other family members or descendants may enjoy learning about your life and experiences very much, and may be influenced to some degree by them. These folks may have had little interaction with us in life, but because of our journals, may come to revere us after we are gone!

In addition to these benefits, after we have journaled for a number of years, it can be fun and provide personal satisfaction when we re-read our journal entries. Sometimes we will pick up on things that we thought or said or did that make us think how we might need to change in the present day. Journal entries of the past reveal how we lived our lives and made decisions and experienced things at that stage of life. It is our own personal history, or as I said before, autobiography. We may be informed by these in such a manner as to be better able to face the days ahead.

Here are some journaling tips. First you would need to have a vision of what you want to write about firmly in mind. Choose your weapon (a computer? A notebook and pen?)

and the time of day to consistently write your thoughts down. If you go digital, there's the benefit of spell checker, as well as legibility. Regarding when to write, the experts say first thing in the morning or at bedtime are typical. As a writer, I get most of my work done in the evening, although truth be told, it is early morning as I am writing this! The other thing I notice is sometime I get ideas at odd times throughout the day. At such times, I often use my i-phone to send messages to myself that captures the content. I have learned over the years that my memories of specific events is not permanent or maintained in exact detail – and can drift. So, by writing things down as close to "as they happen" as possible, I do not have to rely later on a somewhat incomplete and flawed recall process.

What to write about? Just start by writing your thoughts from today, based upon what has happened. I recall that my dad, who was a man of few words, used to write to me on a weekly basis when I was away at college. It was incredibly thoughtful of him, and I looked forward to his notes. They were brief, always included the temperature and weather, made mention of mother and the dog and cats, plus included a whimsical self-portrait of him smoking his cigar! While this is not journaling, it attests to his commitment to writing on a regular basis – and the fact that the reader (me) appreciated his writings. So start that first entry soon, and stick with it – you and your readers won't regret it!

HOW'S YOUR WALKING SPEED?

We are going to briefly examine the impact of lifestyle factors upon aging and longevity in a moment, but first, a specific question to you. How quickly do you walk? If you have embarked on your journey to better strength through the Old Man Muscle workout, chances are you have toned up both your upper body and your legs. The results should be good grip strength, which is associated with longevity, and, barring any pre-existing joint or other lower body issues, good **walking speed**. The latter has been shown by recent research to be linked to not only physical well-being but also brain health – so chalk up another success thanks to your strength training.

▲ **Figure 7.5** Both physical aging and brain aging have been associated with a person's walking speed.

The findings of the research also showed that pre-elderly adults with slow gait speed were succumbing to aging at a faster rate, as evidenced by the apparent age of their face, their grip strength, and various biomarkers including BMI, blood cholesterol, blood pressure and gum health. It was suggested that such adults represented a potential target for interventions to prevent late-life disability and dementia.

HOW'S YOUR GRIP STRENGTH?

We have already discussed walking speed – but what about the predictive ability of hand grip strength in older adults? It is recognized that preserving muscle strength through physical activity and exercise is an important determinant of disease prevention and longevity. In this regard, hand **grip strength** as measured by a device called a dynamometer has been shown to be a reliable and inexpensive index of muscle strength. As such, grip strength can be viewed as a biomarker for biological age and longevity. And it can be an important component of daily functional tasks (e.g. holding onto railings, unscrewing jars, etc.). A 2018 study examined the grip strength of adults aged 40-69 and correlated grip strengths to heart disease, respiratory disease and cancer. As a result, researchers suggested that a grip-strength measurement of less than 57 pounds for men and less than 35 pounds for women was associated with higher overall risk for morbidity (disease) and mortality (death).

▲ **Figure 7.6** We are taught that a firm handshake creates a good impression. However, there are health and longevity implications for grip strength as well. A strong grip may predict a longer life. In addition to the Old Man Muscle deadlift and row exercises, regularly squeezing a tennis ball, performing what is called a farmer's carry with a dumbbell in each hand (e.g. walking purposefully this way back and forth in your driveway), and hanging for as long as you can from a chin-up bar can all help develop your grip strength.

THE POWER OF LIFESTYLE HABITS

It has been a theme elsewhere in this handbook, and we should all keep in mind, that lifestyle behaviors can have a major impact on health and longevity. Their "power" can affect us both positively as well as negatively. For example, the impacts of exercise, nutrition, and non-smoking have been investigated. These have been shown to have somewhat dramatic impact on increasing how long people live. But what about the spiritual component? There is also research looking at this lifestyle factor – and results indicate that people who regularly attend religious services tend to live longer, by about two to three years. However, attending religious services is not necessarily the same thing as being spiritually strong or mature, plus some faith groups may adopt other proven healthy behaviors that contribute to the added years, so we have to be cautious regarding interpretation here. Other factors that need to be investigated relate to investigating different faiths, the benefits of having social interactions within the faith community, and measuring specific health parameters rather than simply looking at years lived. Some research has already examined parameters such as blood pressure and cortisol levels, and have found positive (lowering) effects.

LIFE STYLE FACTORS	IMPACT
Regular Exercise	A LONGER LIFE
A wholesome diet	
No smoking	
Spirituality	

▲ **Table 7.2** The effects of several researched lifestyle habits upon life expectancy. Each factor is believed to contribute 2-3 years or more to it.

By the way, the similar terms lifespan, life expectancy and longevity actually refer to different things. **Life expectancy** – the actual number of years a person lives (and should be able to expect to live) - can be attributed to and impacted by an individual and their personal health history, genetics, and lifestyle. On the other hand, **lifespan** is the maximum number of years a human being has been verified and recorded to live irrespective of lifestyle. **Longevity** has to do with how long a person lives beyond average life expectancy. The gap between life expectancy and lifespan is 122 years (lifespan) minus 79 years (life expectancy), which is a whopping 43 years.

In all of this, it should be emphasized that extending the quantity (number of years lived) is not necessarily the goal, but rather improving the quality of those years. This in fact has been the main suggestion of the Old Man Muscle handbook – that although we may extend our years through lifestyle changes, perhaps more important than this, we can achieve a higher quality of life in our later years. This means that when we combine physical strength training with spiritual, in conjunction with healthy lifestyle behaviors, we can be happier, healthier, more functional, and have no regrets.

Speaking of lifestyle habits or behaviors, these are simply choices that we make (or fail to make). So, they are under our control, and we can choose to make them habitual. Let's now talk about some potential daily habits we can adopt that can have cumulative and positive impacts on our lives. I have tabulated eight of these, and others could certainly be added – these eight get us started.

CHECKLIST OF DAILY GOOD HABITS
Express thanksgiving for life, family and opportunities
Do something good for someone or show new kindness
Start a new positive relationship or repair a negative one
Accept yourself for who you are while adding new strength
Learn a new skill or begin a new hobby
Become resilient: if life hands you a lemon, make lemonade
Set physical and spiritual goals that you can accomplish
Wait on the Lord

▲ **Table 7.3** A sampling of suggested good habits that we can choose to engage in on a daily basis.

I wish to comment on the last one – that of waiting on the Lord, because to many it will be a new concept. This is a spiritual practice that will help strengthen us if and when we are spiritually weak. It is directly taken from Isaiah 40:30-31, " Even the youths shall faint and be weary, and the young men shall utterly fall: But they that wait upon the LORD shall renew their strength; they shall mount up with wings as eagles; they shall run, and not be weary; and they shall walk, and not faint." The key here is that we need to realize that we as older men can renew our strength, even when those around us who are younger are falling to the ground out of weakness. The basis for needing to be strong is a verse we covered before, but I will resurrect it (pun intended) now:

> *"You shall love the Lord your God with all your heart, and with all your soul, and with all your strength, and with all your mind; and your neighbor as yourself." (Luke 10:27)*

We wait upon the Lord, and love Him with all our strength. And He becomes and is our strength. The simplest application of this concept comes when we pray. It has been said that prayer is the river through which God's strength flows into our lives. When we ask something of God in prayer, perhaps to deliver us from a situation, to help us resist temptation, to know which direction to go in life, or perhaps we are praying for someone else - we wait for the Lord to provide an answer. Sometimes, it is provided through the reading and studying of His Word. In other times, another person or situation intervenes. Regardless, we can relax as we put all our hope and trust in Him. We can be patient, courageous, and strong during our waiting – not sitting idly by, but continuing in our daily strength training and daily good habits.

> *God is able to do far more abundantly than all that we ask or think, according to his power at work within us" (Ephesians 3:20).*

When we wait on the Lord, because of who He is and what He has done and will do in the future, we will never be disappointed. Although God may not answer us in our timing or in the way we anticipate, He will accomplish his will and good purposes in our lives when we wait on Him and continue in prayer. If we move outside of these spiritual habits, we miss out on His blessing.

Be still before the Lord and wait patiently for him; those who wait for the Lord shall inherit the land; But for you, O Lord, do I wait; it is you, O Lord my God, who will answer. (Psalm 37:7)

Then too, there is this amazing realization by the Apostle Paul in his own life:

Concerning this thing I pleaded with the Lord three times that it might depart from me. And He said to me, "My grace is sufficient for you, for My strength is made perfect in weakness." Therefore most gladly I will rather boast in my infirmities, that the power of Christ may rest upon me. Therefore I take pleasure in infirmities, in reproaches, in needs, in persecutions, in distresses, for Christ's sake. For when I am weak, then I am strong. 2 Corinthians 12:8-10

Strength in weakness. This is opposite to what the world teaches! When we understand what is at work here, it all comes down to obtaining strength from the Lord in order to help those who are weak. But here is the irony. Unless we first acknowledge our own weakness to God, and seek his strength, we will fail in our attempts to help the weak. Interesting, right? This is why we must first acknowledge our spiritual weakness before we take steps to train, grow, become strong – and then live a life of service to God and other people, which, after all, is the prime directive.

PHYSICAL STRENGTH

+

SPIRITUAL STRENGTH

=

A BETTER FUTURE

OLD MAN MUSCLE

▲ **Figure 7.7** This poster is hanging on a wall in the Old Man Muscle fitness studio.

We can also consider the flip side of good habits, namely bad habits, and what is to be done about them. Certainly we all have them – and they range from the relatively inoffensive to downright hurtful and harmful to us and even those around us.

Writer Darius Foroux asks this question: How serious are you about living a purposeful, useful and meaningful life? And we could extend that to, how serious are you about physical and **spiritual strength training**? How about adopting a heathy lifestyle? In other words, if we simply talk about such things but make little or no effort to implement the changes needed, we are going to fail. And this includes kicking all our bad habits to the curb. The advice that Foroux gives is first to identify all of your bad habits, and then to decide rather than quitting one at a time, do it "cold turkey" – get rid of them all at once. You can do that right now for five seconds – just keep reading from this page. It will consume your effort and attention during those five seconds and leave you no time for your bad habits.

STAR TREK

▲ **Figure 7.8** The classic Star Trek tv show offered interesting plots and socio-political commentary. I suggest that we may extrapolate from one of the episodes when considering the challenge of eliminating bad habits from our daily lives.

See? You can do it. The key is to extend from five seconds to forever. We can do it if we really want to. The key is to start, and then to remain steadfast. I recall one Star Trek episode, I believe it may have been "A Taste of Armageddon", in which Captain Kirk observes that although the human species is war-like and capable of unthinkable savagery, we can each decide that, just today – today, "I will not kill". And his argument is applicable to our situation, in that today can be the day that we discard all of our bad habits and start with a clean slate – just today. And perhaps, after a string of "just todays", we might just stick to it.

CHECKLIST OF DAILY BAD HABITS
Being a stubborn person
Failing to set goals
Ignoring your health and fitness
Not being focused and organized
Being opinionated without being educated
Complaining all the time
Worrying all the time
Overworking, doing everything yourself
Neglecting one's spiritual strenght training

▲ **Table 7.4** It is useful to take a personal inventory of one's so-called bad habits in order to successfully address them. Although not specifically listed here, under health we could expand to: eating fried foods, drinking sugary beverages, overindulging in desserts, smoking, and drinking alcohol in excess.

PREVENTING COGNITIVE DECLINE WITH PHYSICAL ACTIVITY

In case you missed it, back in Chapter 3 we had a section discussing nervous system adaptations to exercise. This included information regarding improving executive function through physical activity. The topic is quite relevant to our population of older men due to the fact that cognitive decline in older adults as a whole is increasing, and is one of the largest factors in the loss of independence. As we age, there is increased risk of developing dementia, the most recognized form being Alzheimer's disease. This is a condition in which cognitive ability progressively degenerates. Memory losses initially occur that are disproportional to the individual's age, and this is termed MCI, **mild cognitive impairment**. As impairments increase in severity, problems arise in executive function. Sufferers experience difficulty in planning and organizing, which in turn limit the ability to complete basic activities of daily living.

The good news is, research has demonstrated that exercise can help maintain both physical and mental health. Specialized exercises that combine physical activity and cognitive activity have been shown to reduce cognitive decline. It is important for trainers as well as clients to recognize the plastic potential of the brain, even in advanced age, such that cognitive decline can be successfully prevented or delayed. Trainers need to become educated regarding how to integrate cognitive tasks into fitness programs. A simple example that I have devised is the following.

(Directions to the trainer): Have your client stand upright, holding a somewhat light pair of dumbbells that he can easily curl 20 times at his side. Explain beforehand that you will be providing verbal instructions to him that require he simultaneously march in place at a specific cadence (speed) while performing whatever dumbbell exercises that you call out. The trainer should decide on a reasonable cadence at first, and may increase it during the exercise.

For example: "Begin marching in place. Perform one set of ten alternating dumbbell curls as you count backward from the number twenty-five." As the client finishes and keeps marching in place, ask, "What number did you end with?" (It should be sixteen if executed properly).

▲ **Figure 7.9** Designing exercises and activities that stimulate the brain and the body simultaneously have been shown to significantly improve parameters of executive function. Such a strategy may serve as an effective primary prevention strategy of cognitive decline.

"Next, while walking carefully between three evenly spaced pylons on the floor, perform three double-arm triceps kickbacks (or you could again use biceps curls, or mix-n-match), one per pylon. As you complete each kickback, shout out the numbers 781, then 187, and lastly, 817." Have the client reverse direction (turn around after the third pylon) and repeat this task two more times in each direction. The trainer should provide different random number sets each time, and try to progressively quicken the pace of the exercise within reason depending upon the age and ability of the client.

As you can tell, these activities will require your client to focus their thinking on several things simultaneously while also engaging muscular contraction, coordination, balance and movement. Such an activity combines physical and cognitive training. Of utmost importance is to watch for the client's safety – you should allow for a slow rehearsal before launching into this activity. Another key point is to keep the activity fun and up-beat. Laughter and encouragement can be essential elements of this kind of training. Never allow the activity to be so overwhelming and difficult that the client becomes frustrated to the point of quitting. If you see that happening, modify the activity so that it is less challenging.

And now, it is time to close out The Old Man Muscle Handbook. You know, it was so much fun to write, and I was able to pour so much diverse information into this handbook, I almost don't want it to end. But in a sense it is only the beginning. And I say this because each of us now has the chance to take some or all of the information presented within these pages and apply it to our life. I hope you have had a good and successful time with it.

If the reader is an older man, then like me he is in the winter season of life. During this time of one's life, the days, weeks, months and years pass by more quickly. And the years that remain become what now count - they are what we have left. How shall we live them? Probably we have realized by now that there are no guarantees in life – we all think that we will wake up tomorrow and do certain things, but this is uncertain. As I believe I said earlier, we all have an expiration date. However, it is my fervent hope and prayer that we all will be granted enough time to live in such a way so as to be a benefit to others. By this I mean those people we know and love of course – but also to those who are mere acquaintances, and even to strangers. In the final analysis, what is more important than the size of our house or bank account or the number of our diplomas - is how we treat people. Someone once said, "People will perhaps forget your name, they will forget what you said, they will forget what you did, but they will NEVER forget how you treated them."

This book has been mainly a "self-help" handbook. It is sort of a combination of books on health, fitness, nutrition, aging, and faith, all in one. However, we should not be so self-focused that we miss one key message. And it is, that life is not all about us. And from what we just discussed, the impact of how we treat people cannot be underestimated. We should consider the wisdom of this teaching. How will you and I be remembered? As a kind, fair and giving person? Or as an ambitious self-promoter who never did anything for anyone but himself? We can make a positive difference in people's lives, and we should try, while we have still been blessed with days on this planet. As we improve ourselves physically and spiritually, it is a noble and worthy sentiment to care more about others than about ourselves, and to display a compassionate and giving spirit.

We can learn from the parable of the Good Samaritan:

> *In reply Jesus said: "A man was going down from Jerusalem to Jericho, when he was attacked by robbers. They stripped him of his clothes, beat him and went away, leaving him half dead. A priest happened to be going down the same road, and when he saw the man, he passed by on the other side. So too, a Levite, when he came to the place and saw him, passed by on the other side. But a Samaritan, as he traveled, came where the man was; and when he saw him, he took pity on him. He went to him and bandaged his wounds, pouring on oil and wine. Then he put the man on his own donkey, brought him to an inn and took care of him. The next day he took out two denarii and gave them to the innkeeper. 'Look after him,' he said, 'and when I return, I will reimburse you for any extra expense you may have.' "Which of these three do you think was a neighbor to the man who fell into the hands of robbers?" The expert in the law replied, "The one who had mercy on him." Jesus told him, "Go and do likewise." (Luke 10:30-37).*

Fight and Protect, or Pray and Serve?

▲ **Figure 7.10** As a species, and a gender, the characteristic of masculinity and the role of the male gender have been predicated upon the belief that our instinct included being ready to fight when confronted by another male, and being ready to protect spouse and family from perceived threats of any kind. In modern society, with a great ratcheting down of such challenging external stimuli, I submit that we make better use of our energies in meeting the prime directive mentioned previously: living for God's glory, and placing the needs of others before self.

Writer Steve Bloom has stated the following: "There's a link between helping others and finding meaning in your life. And it's not just those who have already found purpose giving back. Instead, helping others can actually CREATE a sense of meaning in our lives. Helping others fulfills some of our most basic needs such as connecting with others and seeing how your actions make a positive difference. Those are the areas that help define a meaningful life. A meaningful life isn't found, but created through our actions. And it starts with looking for ways to help others."

What a positive and motivational thought to end this book. A better life at last. From one old man to another: I wish you all the best with your training - and with your new life!

Agronin ME. 2018. The End of Old Age: Living a Longer, More Purposeful Life. Da Capo Lifelong Books.

Basu-Zharku IO. 2011. The Influence of Religion on Health. 3(1):1-3. Accessed at: http://www.inquiriesjournal.com/articles/367/2/the-influence-of-religion-on-health

Bloom S. 2018. 7 Reasons why helping others will help you live a better life. Accessed at: https://dosomethingcool.net/helping-others-life/

Christie GJ, et al. 2017. Do Lifestyle Activities Protect Against Cognitive Decline in Aging? A Review. Front Aging Neurosci 9:381. Accessed at: https://www.ncbi.nlm.nih.gov/pmc/articles/PMC5701915/

Foroux D. 2019. Why you can't break your bad habits. Accessed at: https://getpocket.com/explore/item/why-you-can-t-break-your-bad-habits?utm_source=pocket-newtab

Gheysen F, Poppe L, DeSmet A, et al. 2018. Physical activity to improve cognition in older adults: can physical activity programs enriched with cognitive challenges enhance the effects? A systematic review and meta-analysis. Int J Behav Nutr Phys Act 15, 63. Accessed at: https://ijbnpa.biomedcentral.com/articles/10.1186/s12966-018-0697-x#citeas

Green L. 2017. What to do while you're waiting on God. Accessed at: https://unlockingthebible.org/2017/08/what-to-do-while-youre-waiting-on-god/

Hall DE. 2006. Religious Attendance: More Cost-Effective Than Lipitor? The Journal of the American Board of Family Medicine 19(2):103-109.

Hone L. 2019. Sorrow and tragedy will happen to us all — here are 3 strategies to help you cope. Accessed at: https://ideas.ted.com/sorrow-and-tragedy-will-happen-to-us-all-here-are-3-strategies-to-help-you-cope/?utm_source=pocket-newtab

Kadlec D. 2019. How to reinvent your life after 50. Accessed at: https://www.considerable.com/work/second-careers/reinvent-you-life-after-50/

Liu-Ambrose T, et al. 2010. Resistance training and executive functions: A 12-month randomised controlled trial. Arch Intern Med 170(2): 170–178.

McKay B, McKay K. 2009. 30 Days to a better man Day 8: start journaling. Accessed at: https://www.artofmanliness.com/articles/30-days-to-a-better-man-day-8-start-a-journal/

Mind Tools Content Team. Developing Resilience: Overcoming and Growing From Setbacks. Accessed at: https://www.mindtools.com/pages/article/resilience.htm

Minson M. 2012. Prepare to defend yourself… how to age gracefully & escape with your dignity. Texas A&M University Press.

NIA. 2012. Report: Growing Older in America: The Health and Retirement Study. Accessed at: https://www.nia.nih.gov/sites/default/files/2017-06/health_and_retirement_study_0.pdf

Rasmussen LJH, et al. 2019. Association of neurocognitive and physical function with gait speed in midlife. JAMA Network Open. 2(10):e1913123.

Scripture (bible) verses. Accessed at: www.biblegateway.com

Spector N. 2019. 'OK boomer' is dividing generations. What does it mean? Accessed at: https://www.nbcnews.com/better/lifestyle/ok-boomer-diving-generation-what-does-it-mean-ncna1077261

Stulberg B. 2018. An Ode To Being Old: On the harder-to-measure benefits of age and experience. Accessed at: https://www.outsideonline.com/2339366/benefits-getting-older-health-fitness-aging

Whitney DG and Peterson MD. 2018. The association between differing grip strength measures and mortality and cerebrovascular event in older adults: National Health and Aging Trends Study. Front Physiol 9:1871. Accessed at: https://www.ncbi.nlm.nih.gov/pmc/articles/PMC6330303/

Index

T

V

W

X

Y

actin and myosin the functional contractile proteins of muscle tissue

acute systemic response how the body responds to a single bout of exercise

adaptations specific ways that body systems (neuromuscular, cardiorespiratory, etc.) respond to exercise

adaptogen substances that can alleviate the negative effects of exercise and stress, including fatigue, depression, and anxiety

ADLs (activities of daily living) the range of daily activities that people engage in that requires muscle power, such as getting up from a lying or seated position

aerobic with oxygen; aerobic metabolism uses fat to make ATP energy; an aerobic exercise example is cardio

ageism discrimination directed at a particular age group or generation

aging (senescence) is the sum total of biological and physiological processes that begin to decline beyond adulthood; the accumulation of damage in molecules, cells and tissues over a lifetime that compromises functional ability

AMPK (AMP-activated protein kinase) a primarily catabolic kinase enzyme involved in cell signaling that opposes the anabolic actions of mTOR

anabolic "synthesis"; that which builds up or causes growth

anabolic metabolism the form in which molecules required by the body are built, or synthesized, typically requiring ATP energy input; also refers to reactions that favor tissue (i.e. muscle) building

anabolic resistance blunted response to dietary protein, to exercise, and to muscle protein synthesis that is associated with aging

anabolic steroids man-made analogs of the male sex hormone testosterone, taken in order to promote muscle mass; their unsupervised use is not without risks

anaerobic without oxygen; an anaerobic exercise example is strength training

anaerobic energy is generated via metabolism that uses glucose to make ATP energy

angiotensin-converting enzyme an enzyme that regulates body fluid volumes and pressures; when inhibited, this allows blood vessels to relax and brings down blood pressure

anthocyanin natural plant-derived compounds (pigment molecules) that have a positive impact on neurological processes and may slow the age-related decline in brain and cognitive functions

aromatase enzyme that converts testosterone into estrogen

ashwagandha medicinal herb that may aid in MPS and alleviating stress

astaxanthin pigment molecule in the carotenoid family having both antioxidant and anti-inflammatory properties; marketed as a drug combined with Vitamin D3 as an anti-sarcopenic therapy

ATP (adenosine triphosphate) a high energy molecule generated via energy metabolism and used by body cells, tissues, organs to do work

BDNF (brain derived neurotrophic factor) a specific cytokine in muscle that regulates satellite cell behavior and muscle regeneration; may aid the conversion between type 1 and type II muscle fibers; also present in brain where it acts to enhance neurogenesis and cognitive functioning

behavior goal when setting goals, the behavior is what will get you to your goal

bench press one of the strength training exercises in the Old Man Muscle workout

beta alanine a non-essential amino acid that has been shown to enhance exercise performance, specifically strength gains

blood pressure a measure how hard the blood is pushing against the walls of the arteries, given in systolic pressure (heart pumping)/

diastolic pressure (heart resting) numbers, such as 120/70

BMI (body mass index) an assessment tool to identify underweight, normal weight, overweight, and obesity; the formula is body weight (in kg) divided by height (in meters2)

calf raise one of the warm-up exercises in the Old Man Muscle workout

can opener a dynamic stretch used prior to the Old Man Muscle workout

catabolic "breaking down" as in a catabolic reaction that breaks substrates into products; see also catabolic metabolism

catabolic metabolism the form in which complex molecules like starches, proteins, and triglycerides are broken down into smaller molecules; also refers to reactions that favor tissue (i.e. muscle) breakdown

cholecalciferol technical term for the active form of Vitamin D

concentric muscle contraction during strength training as the weight is lowered and the muscle lengthens

cool-down a low-impact post-workout activity (such as walking) that allows breathing and heart rate to return to their normal pre-workout status, thereby greatly reducing the risk of an adverse cardiac event

cortisol so-called stress hormone (catabolic) produced by adrenal glands that plays a role in glucose metabolism

creatine a supplement marketed as a muscle-building aid that helps generate ATP in working muscle

curcumin a phenol compound which is a component of the spice called turmeric, having both antioxidant and anti-inflammatory properties

curl one of the strength training exercises in the Old Man Muscle workout

dead shoulder circles a dynamic stretch used prior to the Old Man Muscle workout

deadlift one of the strength training exercises in the Old Man Muscle workout

dehydration state of body water depletion

DHEA (dehydroepiandrosterone) a prohormone secreted by the adrenal glands that is enzymatically converted into the primary hormones estrogen and testosterone in the body

DNA methylation an enzymatic process that adds CH_3, (methyl) groups directly to DNA, causing the genes under effect to switch on or off

DOMS (delayed onset muscle soreness) a normal response to overload training especially in a person unused to strength training, caused by muscle tissue damage and probably also inflammation, resulting in post-workout pain and muscle weakness

duality of existence being created with both a physical and spiritual nature

dumbbell swing one of the strength training exercises in the Old Man Muscle workout

DUP daily undulating periodization, a training approach that seeks to understand the impact of different repetition ranges on muscle strength and growth

dynamic stretching a form of stretching that involves body movement, and serves as a good mobility-enhancing prelude to a strength training workout

eccentric muscle contraction during strength training as the weight is raised and the muscle shortens

endocrine disruptor substance such as the plastics that has been shown to disrupt the functioning of the human endocrine (hormonal) system

endocrine system the organs that secrete hormones, which regulate metabolism, cell growth and function, and many other body processes

endurance ability to maintain a specific muscular activity over time, such as when jogging or cycling

engrams (mental engrams) a set of instructions, for instance from brain to working muscle

epicatechin a green tea flavanol that may assist with muscle growth, possibly by inhibiting myostatin

epigenetic modifications enzymatic changes to DNA (i.e. methylation) causing genes to become activated (turned "on") or deactivated (turned "off")

epigenetics the impact of the environment

and lifestyle on body cell activity over time, particularly when it comes to gene expression

EPOC excess post exercise oxygen consumption, referring to the increased rate of oxygen uptake – hence energy expenditure - after working out; "the afterburn effect" in post-exercise calorie burning at rest

excitation–contraction the sequence of events that start with a motor nerve impulse and result in muscle contraction

executive function brain processes that help us manage our day and plan activities

fast twitch type of muscle fiber used for power activities; low oxidative, high glycolytic; "white" fiber

FFM fat-free mass is body composition components other than fat (bone, muscle)

fish oil supplement containing omega-3 fatty acids (DHA and EPA) that act in an anti-inflammatory manner, and is potentially anabolic

food environment the presence and availability of specific foods at a specific location (i.e. the home)

free radical theory of aging proposes that when the body's antioxidative capacity becomes overwhelmed by too many free radicals (causing oxidative stress damage), over a lifetime this leads to aging

glucagon hormone secreted by the alpha cells of the pancreas that acts in opposition to insulin; when blood glucose is low (hypoglycemia), glucagon acts to cause the breakdown of glycogen (stored in muscle and liver) to generate glucose

gluconeogenesis formation or synthesis of glucose from non-carbohydrate sources (i.e. fat and protein breakdown)

glycogenolysis breakdown of glycogen stored in muscle and liver to generate free glucose

goblet squat one of the strength training exercises in the Old Man Muscle workout

GOMS grumpy old man syndrome; grumpiness in older men caused by a variety of physical declines and psychological factors

grip strength a measure of strength and a predictor of longevity, it is a measure of the muscular strength generated by the forearm muscles as measured by a dynamometer

growth hormone (GH or HGH) major anabolic hormone secreted by the anterior pituitary gland; facilitates protein synthesis, nutrient transport, and stimulates cell growth

hamstring reach a dynamic stretch used prior to the Old Man Muscle workout

heart rate the rhythmic contractions of the heart as measured in beats per minute (bpm), as assessed by the pulse or by listening to the heart with a stethoscope

HIIT high-intensity interval training a popular form of intense workouts that combine catabolic and anabolic exercises

histones proteins associated with the DNA within chromosomes

HMB (Beta-hydroxy beta-methylbutyrate) derivative of the amino acid leucine that is thought to be anti-catabolic (prevents muscle breakdown) as well as aid in muscle hypertrophy and strength development

hormones peptide (protein molecule) chemical signals that travel from the organ of their origin to target tissue in order to cause some sort of effect

hypercholesterolemia elevated blood cholersterol levels

hypertension high blood pressure (AHA defines "stage 1" hypertension as values of or greater than 130/80)

hypertrophy an increase in size, as in muscle hypertrophy due to strength training

hypothesis a statement about something that we believe to be true

immunosenescence aging of the immune system, resulting in its functional decline

IMS (irritable male syndrome) medical term for grumpy old man syndrome

Inchworm a dynamic stretch used prior to the Old Man Muscle workout

inflammaging chronic, low-grade (harmful) inflammation associated with aging

inflammation the body's natural response to foreign substances (i.e. bacteria) as well as to cell and tissue damage; can be brief (acute, beneficial) or long-term (chronic, harmful)

insulin hormone secreted by the beta cells of the pancreas that acts in opposition to glucagon; when blood glucose is high (hyperglycemia), insulin acts to clear glucose from the blood by facilitating entry into body cells, where it generates ATP energy

insulin-like growth factor anabolic hormone produced in the liver that supports the function of HGH to repair protein damaged during exercise, and stimulates the proliferation and differentiation of satellite cells

intermittent fasting an modified fasting approach that minimizes the timing of our daily eating periods, to either 10, 8, or 6 hours in a 24-hour period, the remainder of the time we fast and only drink water or unsweet tea

isometric during muscle contraction, a weight is held steady, and often consciously and forcefully flexed, without moving it up, down, or in any direction

journaling the process of writing down personal thoughts, experiences, and observations in an organized fashion and on a daily basis

ketogenic diet a low carbohydrate (30 grams maximum), higher fat and protein diet that restricts sugary, bakery and grain foods items in favor of eggs, meats, seafood, nuts, and dairy

ketones breakdown products of fat metabolism that increase during a keto diet, providing an alternate fuel source to glucose

ketosis a condition wherein the body begins to deplete stored fat, generating a build-up of ketone bodies in the blood

LBM (lean body mass) includes the combined weight of the internal organs, bones, muscles, body water, ligaments and tendons, plus the weight of the essential fat in the organs, central nervous system, and bone marrow

leg lift a hamstring and core rehearsal exercise performed prior to the Old Man Muscle workout

leg press a quad and hamstring rehearsal exercise performed prior to the Old Man Muscle workout

leucine one of the BCAAs (branched chain amino acids) and the primary amino acid to trigger MPS

life expectancy the actual number of years a person lives (and should be able to expect to live) as attributed to and impacted by an individual's personal health history, genetics, and lifestyle

lifespan the maximum number of years a human being has been verified and recorded to live irrespective of lifestyle

lifestyle how we live our lives day in and day out over an extended period of time

lipolysis fat or triglyceride breakdown to yield free fatty acids

longevity how long a person lives beyond average life expectancy

macronutrients the nutrients, including protein, fat and carbohydrate that the body requires in relatively greater amounts (milligram to gram range per day)

magnesium an essential mineral that is important in a great man metabolic processes, it may also be effective in stress reduction

mechanical overload the expected structural damage to muscle fibers as the result of lifting weights during strength training

mental engrams (see engrams) neural networks in the brain that ultimately instruct the muscles and joints how to perform

metabolic health requires optimal levels of blood fats (triglycerides), blood sugar (glucose), blood pressure, high-density lipoprotein cholesterol (HDL), and waist circumference

metabolic overload a condition achieved through metabolic training that forces muscles to fatigue, at which point they generate metabolic products like hydrogen and lactate ions that decrease blood pH and stimulate growth hormone release

metabolic syndrome a cluster of three of more risk factors occurring simultaneously in a person, including increased waist circumference, elevated blood lipids (triglycerides), reduced HDL-cholesterol, high blood pressure, and elevated fasting blood glucose (insulin resistance)

metabolism the chemical reactions that take place in the body that provide for digestion of food, absorption of nutrients, energy production, body growth and development, reproduction, and ultimately, good health and survival

microgreens young or "baby" green plants, similar to but smaller than sprouts, often more nutrient dense and richer in plant bioactive molecules than their full-sized counterparts; marketed as a superfood

micronutrients the nutrients such as the minerals calcium and iron and vitamins such as B12 and C that the body requires in relatively lesser amounts (microgram to milligram range per day)

mild cognitive impairment MCI early stage of dementia in which memory losses occur that are disproportionate to the individual's age

mind: muscle connection conscious (mind controlled) and deliberate (pre-thought; intentional) muscle contraction during strength training

mindfulness paying attention to what you are doing, be it taking care to have correct form when exercising or making healthy food choices when eating

Mitochondrial Free Radical Theory of Aging (MFRTA) proposes that an accumulation of too many free radicals (causing oxidative stress damage) in the mitochondria of cells occurs, over a lifetime this leads to aging

moringa Moringa oleifera a healthful plant that exhibits both antioxidant and anti-inflammatory properties when consumed as a tea or as ground leaves

motor units connections between the central nervous system and individual muscle fibers, in which a single neuron typically innervates multiple muscle fibers

MPS muscle protein synthesis, the formation of new muscle tissue involving satellite cell activation

MRFs (myogenic regulatory factors: MyoD, Myf5, Mrf4, and myogenin) are transcription factors that function to ensure that quiescent satellite cells undergo sequential and complete differentiation into new myofibers

mTOR a kinase enzyme specifically active in skeletal muscle as part of the cell signaling process that responds to nutrients (protein intake) and helps the body build muscle; the acronym stands for mechanistic target of rapamycin

muscle fatigue a positive goal in strength training when seeking to perform reps and sets to failure, but also a negative end product of overtraining and insufficient rest between workouts that can lead to pain, injury and sub-par workouts

muscle fibers long, cylindrically shaped skeletal muscle cells; myofibers

muscle overload subjecting muscles to progressively heavier training loads

myogenesis the formation of new muscle, involving the differentiation and maturation of satellite cells into mature myofibers

myoglobinemia uptake of muscle myoglobin into the blood resulting in elevated levels

myostatin a muscle cell cytokine that acts as a muscle growth inhibitor (catabolic factor) that creates effects opposite to mTOR

neurogenesis formation of new neural connections, including new neurons (i.e. brain cells)

neuromuscular control the communication between the brain and muscles, a critical relationship that influences all variables of muscular fitness

neuroplasticity the ability of the brain to change itself, such as remodel via neurogenesis

nitric oxide signaling molecule and stimulator of muscle mass, function, and regeneration

NSAID (non-steroidal anti-inflammatory drug) acetaminophen, ibuprofen and aspirin, are popular over-the-counter treatment strategies to address exercise-induced inflammation and pain

nutrient timing the notion that intake of food and water at specific time frames before, during, and after a workout each convey certain benefits or advantages when strength training

OK Boomer a derogatory and dismissive statement by younger generations directed at Baby Boomers

oleic acid a monounsaturated fatty acid that increases HDL (good cholesterol), and lowers LDL (bad cholesterol

one RM (1-RM) one repetition maximum; the amount of weight that can only be lifted with correct form and full range of motion a single time (one rep)

osteopenia refers to more than average bone mineral loss but without the loss to bone density seen in osteoporosis

osteoporosis severe bone demineralization (i.e. calcium loss); literally a condition of "porous bones" that are subject to fracture

outcome goal when setting goals, the outcome is what you want to have happen as a result of achieving the goal

overload principle in order to stimulate muscle growth, it is necessary to apply a physical stimulus at an intensity greater than the body is accustomed to receiving

overtraining engaging in too much exercise relative to rest and recovery periods, resulting in muscle soreness, fatigue, decreased performance, and lack of growth

oxidative stress damaging effects of oxygen on body cells and key molecules that are linked to body metabolism

paradox of aging the fact that as many people age, although at first they often become grumpy, in general this eventually gives way to happiness

performance goal something achieved, or the end desired result, like weight loss, or a specific resistance lifted on a strength machine; also called "product goals"; part of SMART goal setting

phytochemicals naturally occurring plant-derived molecules that are not proteins, carbohydrates, fats, vitamins, or minerals but have been shown through scientific research to be of potential benefit to human health

prime mover so-called agonist muscle that is working rather than at rest

pre-, peri-, and post-workout: nutrition the meals or snacks consumed before, during, or after a workout, respectively

process goal something a client does (a behavior) to achieve the performance goal (i.e. "I will eat a salad every day"; "I want to lose ten pounds"); part of SMART goal setting

pronated grip a palms down grip when using barbells or dumbbells

pull in (seated row) one of the strength training exercises in the Old Man Muscle workout

purposeful nutrition thinking about the consequences of what we eat and drink in terms of effects on health

QoL (quality of life) an umbrella term that suggests life quality markers in terms of health status and functioning, life satisfaction and fulfillment, as well as social and other parameters important to an individual

quercetin a bioflavonoid molecule reported to have anti-inflammatory, antioxidant, anticancer, antihypertensive, and blood glucose-lowering properties

reinvention the process in which we recognize the changes happening to us and around us, which cause us to think critically about, adapt, and adjust the direction of our career and life

resilience the ability to regroup and get back on track with our lives in spite of adversity

resting heart rate (RHR) the number of times the heart beats per minute at rest (typically between 60-100)

rhabdomyolysis a proposed cause of DOMS in which stress-related metabolites and by-products of tissue damage create post-workout pain

RICE standard treatment for soft tissue injury that stands for rest or restricted activity, ice application to reduce swelling and inflammation, compression, and elevation

risk stratification an important health assessment activity conducted by personal trainers with clients to identify cardiovascular disease risk based upon family history and other factors

ROS reactive oxygen species (free radicals) that include the superoxide ion (-O2•), the hydroxyl radical (•OH), and the peroxide radical (-O2•)

sarcomere the basic functional unit of muscle

sarcopenia progressive and potentially debilitating loss of muscle size, strength, and function as we age

sarcoplasm a liquid component of muscle cells similar to the cytoplasm of other cells

satellite cells are a type of stem cell in muscle tissue that can be stimulated to differentiate into mature muscle cells (myogenesis), creating proliferation and hypertrophy

SCA (sudden cardiac arrest) is when the heart muscle stops beating; in the context of training, there is greater risk of this happening when a person who trains intensely does not allow the heart rate to decrease to normal by incorporating a proper cool-down after their workout

sinensetin a plant flavonoid (abbreviated as SIN) shown to reverse sarcopenia and boost satellite cell function in rat studies

single arm row one of the strength training exercises in the Old Man Muscle workout

sleep cycle a repeated period of approximately 90 minutes, each comprised of stages that include pre-sleep, non-REM light sleep, non-REM deep sleep, and REM deep sleep

slow twitch type of muscle fiber used for endurance activities; high oxidative, low glycolytic; "red" fiber

SMART goal is a goal that is S (Specific), M (Measurable), A (Attainable), R (Relevant) and T (Time bound)

spiritual strength training embarking on specific activities such as bible study in order to develop the spiritual side of self

static stretching a form of stretching that follows a strength training workout and requires holding a body position for between 15 and 30 seconds

strength the ability to create maximal force during a single contraction of a muscle

strength training (resistance training) is training the body in order to increase skeletal muscle strength; the muscles are forced to work against and overcome resistance loads, such as weights

supinated grip a palms up grip when using barbells or dumbbells

synapse a junction between cell types, for instance, neurons and muscle cells

target heart rate the recommended or desirable heart rate to achieve during exercise, which varies according to age and fitness level

telomeres the end "cap" regions of chromosomes (composed of bundled DNA) that shorten with age

testosterone the male sex steroid hormone; following strength training it is involved in the activation of satellite cells and what is called "myonuclear accretion", enhancing muscle protein synthesis and hypertrophy

testosterone, albumin-bound is a biologically unavailable, inactive form of this hormone bound to albumin in the blood circulation

testosterone, free bioactive form of testosterone that acts as a major body anabolic hormone

testosterone, SHBG-bound (sex hormone-binding globulin) is a biologically unavailable, inactive form of this hormone bound to glycoprotein

testosterone, total combination of the various forms of inactive and active testosterone in the blood circulation

theanine naturally occurring green tea amino acid reported to produce calming, anti-stress effects

thruster one of the strength training exercises in the Old Man Muscle workout

time under tension (TUT) intentionally controlled duration of muscle contraction during an exercise in order to activate more motor units and potentially create metabolic products that contribute to growth

training effect the result of various body adaptations to strength training that typically result in muscle hypertrophy and strength gains

training volume the sum total of three mathematical products: (amount of weight lifted) x (number of repetitions) x (number of sets) in your workout

transcription factors substances that function to make certain that the correct genes are expressed in specific cells of the body at the right time

trapezius shrug one of the strength training exercises in the Old Man Muscle workout

TRF (time restricted feeding) the approach used in animal studies that essentially means the same as intermittent fasting, in which the daily feeding window is restricted in order to generate metabolic health benefits

triceps kickback one of the strength training exercises in the Old Man Muscle workout

Type I error occurs when we reject a null hypothesis that is actually true

Type II error occurs when we fail to reject a null hypothesis that is not true

ventilation the act of inhaling (inspiration of air with oxygen into the lungs) and exhaling (expiration of carbon dioxide out from the lungs)

ventilatory threshold the point at which breathing becomes labored during exercise; subdivided into VT1 and VT2

vertical alignment visual head-to-toe assessment of upright body static posture

Vitamin D (cholecalciferol) a fat-soluble vitamin that also acts as a hormone and is anabolic in Vitamin D deficient individuals

VMS (vanishing muscle syndrome) the loss of observable muscle mass due to lack of training, due to the fact that "muscle does not last" without regular workouts

VO_2 max is the best indicator of cardiovascular endurance; it measures the maximal volume of oxygen that a person can use per minute per kilogram of body weight

VT1 ventilatory threshold 1 is the first ventilatory threshold, the point during exercise at which the breathing rate begins to increase, lactate begins to accumulate in the blood, and speech becomes uncomfortable

VT2 ventilatory threshold 2 is the second ventilatory threshold, and is an indicator of greater exercise intensity with more lactate build up in the blood than at VT1; conversation becomes impossible at this stage

warm-up specific pre-workout activities that serve to increase heart rate and blood flow to slightly raise the temperature of muscle so as to decrease injury risk

whey protein one of the two major proteins in milk, in addition to casein; it is the food protein with the highest biological value, even exceeding that of egg protein

xenoestrogens substances, including endocrine disruptors, that mimic the female hormone estrogen and cause feminizing and sexual dysfunction effects in males

The coronavirus COVID-19 outbreak in late 2019/early 2020 created an unanticipated global pandemic event that caused disease and death, disrupting economies and societies all over the world. It is expected to change our lives in many ways for the foreseeable future. For the fitness industry, it meant that social distancing and stay-home guidance created a barrier to training at fitness centers, as well as with personal trainers in the home. Individuals either had to learn to train themselves, or go online for instruction, using their own home-based gym or equipment.

An important point that I wish to make involves the role of exercise on the immune system and in inflammation. This is relevant because it is the immune system that protects the body from foreign invaders such as bacteria, parasites – and viruses. Events such as bodily injury, microscopic invaders - and even exercise – generate inflammation. That's right, exercise is known to be an inflammatory process. So, we might think that it is a bad thing. Although exercise is a form of stress and does increase acute inflammation, this form of response via exercise is desirable. It is the other kind of inflammation – chronic – that is undesirable. In the case of a virus, in addition to evoking an inflammatory response, harmful effects are generated that can impact multiple bodily organs and systems that can be lethal.

▲ After a moderately intense workout, the following should be the case: (1) you feel like you've worked hard but not like you've been run over by a truck; (2) you feel invigorated, better than before you began the exercises; (3) after you cool down, you feel you could do it all over again. On the other hand, if you feel wasted, the intensity level was too high. The point is, when these criteria are met, you are at the right intensity level; and will be able to recover, even at an advanced age, quickly - and be ready for the next one. You will notice as the workout ends, that due to your elevated heart rate and ventilation (respiration rate/ breathing) throughout the workout, your brain has been invigorated (greater blood movement with oxygen and nutrient delivery, as well as waste removal) and you feel mentally sharper. I get some of my best thinking done at such times, due to enhanced clarity of thought.

Regular exercise helps the immune system to be in tip-top form to mount a strong defense against invaders by reducing chronic inflammation. This is because if the immune system is continually battling chronic inflammation due to the presence of a poor lifestyle, it's attention will be diverted, and its effectiveness lessened - when faced with new challenges – such as a virus. Recall that an unhealthy diet can be pro-inflammatory and can contribute to chronic inflammation, as can being overweight or obese. So if the goal is to maintain a strong immune system during the challenge of COVID-19, we must not only consider the role of sensible exercise (meaning: regular but not too intense), we must also resist the urge to eat that sweet and fried stuff that will spike chronic inflammation and negate our efforts. A healthy diet abounding in plant-based foods, healthy meats and dairy for high quality protein, and fatty fish, nuts, seeds, whole grains and legumes – is an anti-inflammatory diet. Regarding supplements, in Chapter 5 we identified several that are categorized as "anti-inflammatory". While these may help support immune system functioning, it must be noted that no data yet exists to support their effectiveness against COVID-19 via clinical trials.

As usual, adequate rest, sleep, and hydration are most vitally important – to keep the immune system in top form so that antibodies and so forth can do their jobs. Lastly, I will suggest something else: sunshine. The Vitamin D generated on the skin for the body, by the sun's ultraviolet rays, is anti-inflammatory. So, this vitamin helps reduce chronic inflammation, which helps maintain a strong immune response. Actually, sunshine does more - offering a one-two punch. How? Not only does sunshine increase our Vitamin D, which is good for immunity, but recent studies also show that the coronavirus is fragile in the environment, and is quickly and easily killed by the sun's UV-rays in the air and also on surfaces. Sunshine destroys the viral RNA and **DNA**, which causes the virus to die. So if the weather is good, take your exercise outdoors and get the combined benefit of the workout plus the sunshine!

Be strong and be safe, my friends!

Dr. Peter Murano

Old Man Muscle

May, 2020

OLD MAN MUSCLE About the AUTHOR

Peter S. Murano earned the M.S. and Ph.D. degrees in Human Nutrition and Foods from Virginia Tech in 1986 and 1989, respectively. He is a baby boomer, born in 1954.

He served as Deputy Administrator at the USDA Food and Nutrition Service, Washington, D.C. from 2001-2004 during the George W. Bush administration. In that capacity, he supervised a dozen Federal nutrition assistance programs, most of which developed anti-obesity goals and objectives. He and his FNS Child Nutrition Division team created the HealthierUS School Challenge (HUSSC) in 2003, a voluntary certification initiative recognizing schools enrolled in USDA's Team Nutrition that have created healthier school environments through promotion of nutrition and physical activity. In 2004, he received the U.S. Secretary of Agriculture's Honor Award for Superior Service for implementing the National Fruit and Vegetable Pilot in five states and Indian Tribal Organizations, which has since become a permanent and expanding program for school children.

At Texas A&M University since 1995, Dr. Murano serves as a research and teaching professor in the Department of Nutrition and Food Science, and has published in a variety of academic journals including the Journal of Food Science, Journal of the International Society of Sports Nutrition, Medicine and Science in Sports and Exercise, Advances in Medical Education and Practice, Obesity Science and Practice, FASEB Journal, and Nutrients. He founded and directed the Institute for Obesity Research and Program Evaluation at Texas A&M from 2007-2013. In 2014 he was appointed Vice-Chair of the Texas State Leadership Council for Obesity Prevention.

In 2019, Dr. Murano achieved certification as a personal fitness trainer through ACE (American Council on Exercise) and ISSA (International Sports Sciences Association). As owner and operator of the Old Man Muscle Fitness Studio, he is serious about helping men aged 50 and up get back into shape and build up strength in order to succeed in activities of daily life, combat aging, and remain independent for as long as possible. Using proven old school methods, he devised a brief but effective age-appropriate whole body workout for older men that is performed three times per week. He started using it at age 63 in mid-2017 after four decades of sedentary life, and has transformed his body and health status. Peter is also a firm believer in the value of being strong spiritually, and leads a weekly bible study.

▲ The author relaxing at his desk, with vintage photos of Charles Atlas, Steve Reeves, and Rick Drasin's hand-drawn Gold's Gym logo in the background. Is that really a retro telephone I am using?

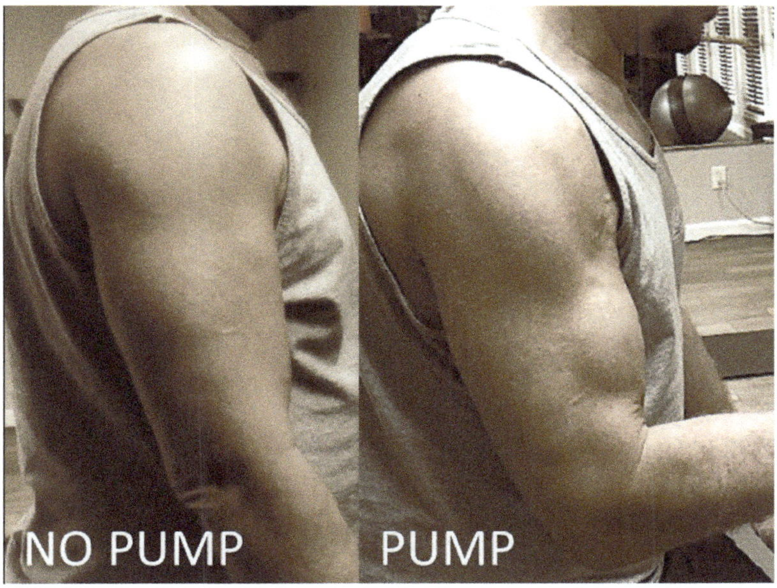

▲ A "muscle pump" can occur – but doesn't always - during training. It happens when the muscles fill more than usual with blood. Nutrients are carried in, and waste is moved out. It only lasts a few moments, but it feels and looks good. New research suggests that aiming for a pump by workout's end may add a growth stimulus to your training. It is gratifying to know that even as I approach age seventy, it is possible for me to get pumped when working out.

HER BOOKS BY THE AUTHOR

THE GEORGE REEVES
ADVENTURES OF SUPERMAN
COMPANION

EXPLORING
FOOD SCIENCE
AND TECHNOLOGY

OBESITY:
A SOCIAL AND SCIENTIFIC
CHALLENGE

▲ In addition to his Old Man Muscle handbook, Peter has authored a tribute book to the 1950s television program, "The Adventures of Superman". In his capacity as an academic writer of textbooks, he is sole author of "Exploring Food Science and Technology" (2nd Edition 2013), and "Obesity: A Social and Scientific Challenge" (2016).